COHERENT QUANTUM OPTICS AND TECHNOLOGY

Advances in Optoelectronics (ADOP)

ADOP Advances in Optoelectronics

COHERENT QUANTUM OPTICS AND TECHNOLOGY

Motoichi OHTSU
Tokyo Institute of Technology, Yokohama

Springer-Science+Business Media, B.V.

Library of Congress Cataloging in Publication Data

CIP DATA APPEARS ON SEPARATE CARD

ISBN 978-94-010-4712-8 ISBN 978-94-011-1632-9 (eBook)
DOI 10.1007/978-94-011-1632-9

Originally published by Kluwer Academic Publishers in 1992
Softcover reprint of the hardcover 1st edition 1992

Preface

While preparing the manuscript of this book, the author was informed that Profs. H. Dehmelt, Univ. of Washington, W. Paul, Univ. Bonn and N. F. Ramsey, Harvard Univ. have won the Nobel Prize in Physics (1989; for their Nobel lectures, refer to Rev. Modern Physics, Vol. 62 (1990) pp. 525–552). Their main contributions are the pioneering works on the Ramsey resonance method, ion trap, and so on, which are the main topics to be discussed in this book.

There are two other groups previously awarded with the Nobel prize in Physics for their contributions to laser physics and quantum optics; C. H. Townes, N. G. Basov, and A. M. Prokhorov were awarded a Nobel prize for their works on the invention of masers and lasers as well as the basic studies of quantum electronics; N. Bloembergen and A. L. Schawlow, in 1981, for their works on laser spectroscopy. It is an exceptional case that a branch of physics could win the Nobel prize three times. This fact reflects and implies that laser physics and physics using lasers are important fields of basic study, which relate closely to the quantum optics and atomic physics. Furthermore, as the case of semiconductor lasers, it has become an important field for industrial applications, e.g., optical communication and optical disk memories. As shown by these examples, it can be considered that the study of laser is a universal one in science and technology, covering a wide range from basic physics to industrial applications.

Studies in these fields exhibited a rapid progress since the invention of the laser in 1960. The theory of quantum optics was developed in the 1960' and 1970's. On the basis of this theoretical study, in the 1980's, laser devices have been drastically improved in performance and thereby spreading industrial applications in a wide variety. These have been also supported by the progress in fabrication technologies of semiconductor laser devices, control and manipulation technologies of atoms and photons, and semiconductor laser-pumped solid-state laser technology. Based on this progress, it has become possible to produce, study, and use the ultimate status of atmos and photons. Furthermore, ultrahigh coherent lasers have been achieved and their various applicaiton systems have been proposed, as will be demonstrated in this book. In this sense, the present status of the study is in the transient stage, i.e., a period of developing towards the maturity in the 21st century. It can be up-to-date to publish this book under such as exciting situation.

Based on the background described above, "COHERENT QUANTUM OPTICS AND TECHNOLOGY", the title of this book, can be defined as a field of study to investigate the principles and methods of generating a very smooth lightwave with low fluctuations, controlling and manipulating atoms and photons and their application systems. Furthermore, generation of non-classical photons by this lightwave is also included in this field. Chapters 1–5 of book have the form of a tutorial textbook for undergraduate students, graduate students and junior engineers, Chapters 6–8 are devoted to introduce the topics still under intensive investigation for the readers to enjoy and appreciate the study of rapidly progressing fields. The author will be happy if the readers are stimulated by these topics and get great achievements in the study of this field. Comments and criticism from the readers with regard to the contents of this book are welcome and deeply appreciated.

The author appreciates Prof. T. Okoshi for providing him with an opportunity of publishing this book. He also thanks Prof. T. Tako (Tokyo Science University), Dr. Y. Suematsu (President, Tokyo Institute of Technology), Profs. K. Iga and M. Ikeda (Tokyo Institute of Technology) for their encouragements to the author's works, Drs. K. Nakagawa and C. -H. Shin, and graduate students of the author's research group for their active supports and discussions, and W. Wang for his critical comments and check of the manuscript.

January 15, 1991

M. Ohtsu

Contents

Chapter 1

Introduction

It has been said that the 20th century was the era of electronics. Since the advent of transistors in 1948, progress in the technical development of fast and large capacity computers, communications, energy storage and transmission, has advanced rapidly owing to the usage of fast integrated circuits developed by controlling the motion of electrons in semiconductor materials. These technologies have been well developed and are promising further potential progress; however, some of the problems yet to be solved are heat dissipation, speed limit of signal transmission and cross-talk among signal channels.

These problems can be solved by using photonics technology in which the motion of photons is controlled and utilized. When using light as waves, this technology is called "lightwave technology" and it can be called "opto-electronics" when the interaction between electrons and light is taken into account. Since the name of this type of technology has not yet been fixed, it will be tentatively called the "coherent quantum optics and technology" in the course of this book. Progress in coherent quantum optics and technology has advanced rapidly since the advent of lasers in 1960. However, because of the 12-year lag behind the birth of transistors, it can be expected photonics will thoroughly have matured by the end of the 20th century. Although there are several problems remaining to be solved in energy storage and transmission, photonics could be fully utilized in the field of computers and communications in the near future. That is, the 21st century will be the era of coherent quantum optics and technology. In prospect of the forthcoming 21st century, or the era of coherent quantum optics and technology, this book intends to review the principles and technology involved in controlling lasers, which is the key device of photonics, in order to generate a very high coherent light (i.e. a smooth lightwave with low fluctuations). Furthermore, advanced technologies, such as the generation of nonclassical photons as well as the manipulation and control of atoms and photon will also be introduced. Since semiconductor lasers are the most promising type of lasers for practical applications, they will be particularly emphasized in this review. However, the topics described in this book can be, in principle, applied to all types of lasers.

1.1 What is a Coherent Light?

The word "coherence" has been widely used not only to describe light but also in many other fields; its most popular meaning is "the similarity among properties of two or more quantities (e.g., materials, signals, informations, and so on)" or, in other words, "cross-correlation among these properties". If the discussion is limited to light, this word can be used to express two meanings. To explain these, the electric field vector is expressed as

$$E(\boldsymbol{r},t) = E_0\cos(\boldsymbol{kr} - \omega t), \tag{1-1}$$

where E_0 is the amplitude, $\boldsymbol{k}$ the wave vector, $\boldsymbol{r}$ the spatial coordinate vector, t the time, and ω is the angular frequency.

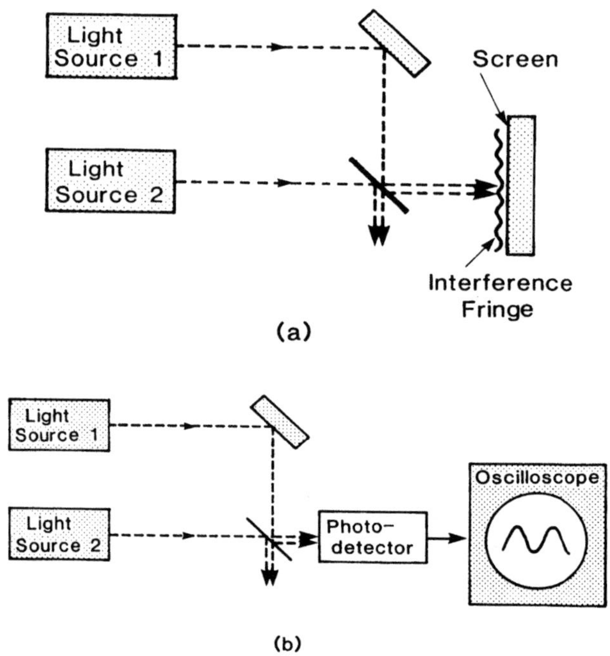

Fig. 1.1. Schematic explanation of the concept of coherence. (a) Spatial coherence. (b) Temporal coherence.

The first meaning of the word is "spatial coherence", which is related to the term $\boldsymbol{kr}$ in this equation. This means, as shown in Fig. 1-1(a), that interference can arise and generate clear interference fringes on the screen when two lights from two light sources (or from one light source which is, however, spatially separated) are superposed. It implies that spatial distributions of the wavefronts of these two lights are similar to each other. Especially, as in the case of light originating from one light source only, it implies that spatial fluctuations of the wavefront are small.

The second meaning of the word is "temporal coherence", which is related to the term ωt in Eq.(1-1). This means, as shown in Fig. 1-1(b), that the clear sinusoidally oscillating beat signal between the two lights can be generated as an output signal from the photodetector. It implies that the temporal behaviors of the two lights are similar to each other. If these two lights are obtained from a common light source, it means that the frequency fluctuation of the light from that light source are very small.

Based on the simple discussions given above, one can easily notice that it is possible to generate a spatially incoherent but temporally coherent light. For example, spatial coherence could be disturbed if light was transmitted through a piece of thin ground glass plate; however, the temporal coherence could still be maintained.

From the above explanations, it can be interpreted that coherent light is a smooth, sinusoidally oscillating lightwave. The discussions in this book will refer to the second meaning, or "temporally coherent light". Such coherent light will be generated by laser, with which a wide variety of applications can be realized, as shown in Fig. 1-2. They range from technological applications such as in communications, sensing, and so on, to scientific applications such as quantum optics, atomic physics, and so on. Details of these application systems will be described in chapter 4.

1.2 Definition and Characteristics of Coherent Light

The explanation of coherent light given in 1.1 was simple and did not give its strict definition. Furthermore, the light emitted from a conventional laser is not always the strictly defined coherent light. In this section, a strict definition of coherent light is given.

The coherent state of light can be defined through the quantization of light [A-D]. For this quantization, one should note that the light can be formulated by Maxwell's equations given by [E]

$$\nabla \times \boldsymbol{H} = \partial \boldsymbol{D} / \partial t \qquad (1-2)$$

$$\nabla \times \boldsymbol{E} = -\partial \boldsymbol{B} / \partial t \qquad (1-3)$$

$$\nabla \cdot \boldsymbol{B} = 0 \qquad (1-4)$$

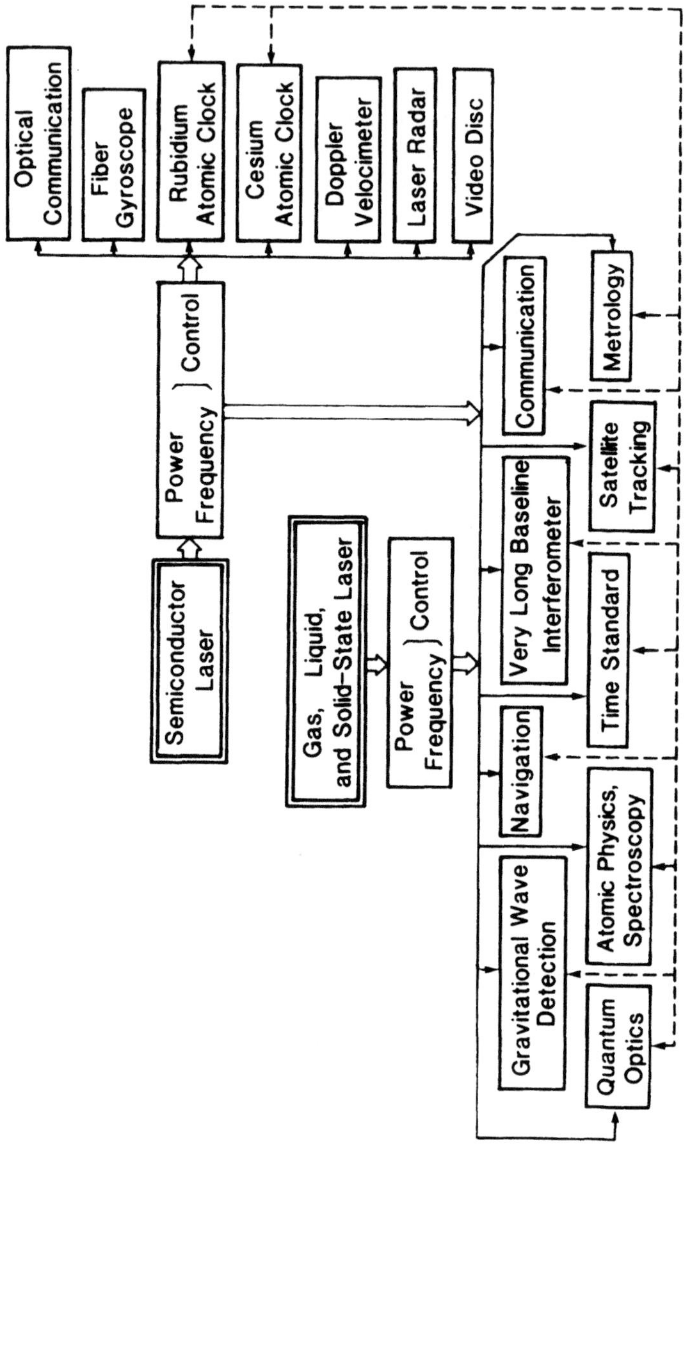

Fig. 1.2. Varieties of the application systems of frequency controlled lasers.

$$\nabla \cdot \boldsymbol{D} = 0 \qquad (1-5)$$

$$\boldsymbol{B} = \mu_0 \boldsymbol{H} \qquad (1-6)$$

and

$$\boldsymbol{D} = \varepsilon_0 \boldsymbol{E} \qquad (1-7)$$

where ***H***, ***E***, ***B***, ***D***, μ_0, and ε_0 represent, magnetic field vector, electric field vector, magnetic induction vector, dielectric displacement vector, permeability of vacuum, and dielectric constant of vacuum, respectively. To discuss the light field inside a laser cavity, it is assumed that light is confined in the cavity shown in Fig. 1-3 and that the wave vector ***k*** is parallel to the z-axis, where Cartesian coordinates are defined by Fig. 1-3. Under this configuration, it is easily found that the electric field

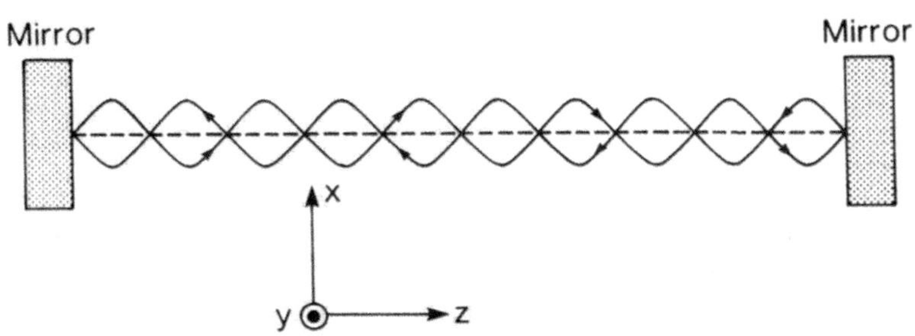

Fig. 1.3. Spatial distribution of the electric field in a cavity with two reflecting mirrors. This figure shows that a standing wave exists which is composed of the two running waves along the +z and –z axes.

vector ***E*** of a plane-wave, linearly polarized along the x-axis and with an angular frequency of ω, is the solution of Maxwell's equations of Eqs.(1-2) – (1-7). The x-component of this vector ***E*** can be expressed as

$$E_x(z\ t) = q(t)\sqrt{\frac{2\omega^2 M}{\varepsilon_0 V}}\sin kz \qquad (1-8)$$

where V is the cavity volume, M is the quantity with the dimension of mass, and $q(t)$ is the quantity with the dimension of length. By using Eqs.(1-1), (1-7) and (1-8), the non-zero component of the magnetic field vector ***H*** is given by

$$H_y(z,t) = \frac{\mathrm{d}q}{\mathrm{d}t}\frac{\varepsilon_0}{k}\sqrt{\frac{2\omega^2 M}{\varepsilon_0 V}}\cos kz. \qquad (1-9)$$

The Hamiltonian of this light is given by

$$\mathcal{H} = \frac{1}{2}\int_V \left(\varepsilon_0 E_x^2 + \mu_0 H_y^2\right)\mathrm{d}v. \qquad (1-10)$$

The integration is carried out for the whole cavity volume. By substituting Eqs.(1-8) and (1-9) into this equation, one obtains

$$\mathcal{H} = \frac{1}{2}\left(M\omega^2 q^2 + \frac{p^2}{M}\right), \qquad (1-11)$$

where the quantity p is defined as

$$p \equiv M\frac{\mathrm{d}q}{\mathrm{d}t}. \qquad (1-12)$$

It is easily found from Eqs.(1-11) and (1-12) that $\mathcal{H}$ has the same form as the Hamiltonian of a harmonic oscillator with a mass of M, an angular frequency of ω, a position of $q(t)$ and a momentum of $p(t)$. Therefore, the quantization of light can also be carried out by following the same procedure as that for the harmonic oscillator, i.e. the q and p are treated as quantum mechanical operators which follow the commutation relation given by

$$[q, p] \equiv qp - pq = i\hbar, \qquad (1-13)$$

where [ ] is the commutator and $\hbar$ is Planck's constant divided by 2π ($\hbar \equiv h/2\pi$). Two alternative operators are defined by

$$a \equiv \frac{1}{\sqrt{2M\hbar\omega}}(M\omega q + ip), \qquad (1-14)$$

and

$$a^+ \equiv \frac{1}{\sqrt{2M\hbar\omega}}(M\omega q - ip). \qquad (1-15)$$

It should be noted from these definitions that a and a^+ are not Hermitian operators. However, the Hamiltonian of Eq.(1-11) is an Hermitian operator which can be expressed by a and a^+ as

$$\mathcal{H} = \hbar\omega\left(a^+ a + \frac{1}{2}\right). \qquad (1-16)$$

The commutation relation of Eq.(1-13) is given by

$$\left[a, a^+\right] = 1, \qquad (1-17)$$

and furthermore,

$$[a, a] = \left[a^+, a^+\right] = 0. \qquad (1-18)$$

The E_x of Eq.(1-8) is then expressed as

$$E_x(z,t) = E_{x0}\left(a + a^+\right)\sin kz, \qquad (1-19)$$

where

$$E_{xo} \equiv \sqrt{\frac{\hbar\omega}{\varepsilon_0 V}}. \qquad (1-20)$$

The operator E_x given by Eq.(1-19) is also an Hermitian operator. Furthermore,

$$[\mathcal{H}, a] = -\hbar\omega a \quad \text{and} \quad \left[\mathcal{H}, a^+\right] = \hbar\omega a^+. \qquad (1-21)$$

The Heisenberg equation of motion for the operator a is

$$\frac{\mathrm{d}}{\mathrm{d}t} a(t) = \frac{i}{\hbar}[\mathcal{H}, a] = -i\omega a, \qquad (1-22)$$

which has a solution expressed as

$$a(t) = a(0)\exp(-i\omega t). \qquad (1-23)$$

A similar procedure gives the expression for $a^+(t)$, which is

$$a^+(t) = a^+(0)\exp(i\omega t). \qquad (1-24)$$

By substituting Eqs.(1-23) and (1-24) into Eq.(1-19) and by noting that $\sin kz = [\exp(ikz) - \exp(-ikz)]/2i$, it can be found that the $E_x(z, t)$ of Eq.(1-19) is composed of $\exp[i(\omega t - kz)]$, $\exp[i(\omega t + kz)]$ and of their respective complex conjugates. From a classical viewpoint, this implies that these exponential terms correspond to the running wave propagating along the $+z$ and $-z$ axes in Fig. 1-3, respectively, and that the standing wave in the cavity is formed by the superposition of these two running waves. Furthermore, it is found that the quantity ω corresponds to the angular frequency of the light whose Hamiltonian is given by Eq.(1-16).

For the discussions given below, the equation for the eigenvalue of the energy W_n and eigenstate $|n\rangle$

$$\mathcal{H}|n\rangle = W_n|n\rangle \qquad (1-25)$$

is solved here. Here, $|n\rangle$ represents the ket vector, which is a vector representation of the wave function appearing in the time-independent Schrödinger equation. By using Eq.(1-21), one can find that

$$\mathcal{H}\, a|n\rangle = (a\mathcal{H} - \mathcal{H}\, a)|n\rangle = (W_n - \hbar\omega)a|n\rangle. \qquad (1-26)$$

This equation implies that the vector $a|n\rangle$ represents the energy eigen-state with the eigen-value of $W_n - \hbar\omega$. Therefore, it is found that the operator a works in decreasing the energy by $\hbar\omega$, from which this operator a is called an annihilation operator. On the other hand, a^+ is called a creation operator because it works in increasing the energy by $\hbar\omega$.

It can be found that the minimum value of W_n takes a positive value. To prove this fact, one should note that the expectation value of an Hamiltonian operator for the arbitrary state $|\phi\rangle$ can be expressed as

$$\langle\phi|\mathcal{H}|\phi\rangle = \hbar\omega\langle\phi'|\phi'\rangle + \frac{1}{2}\hbar\omega, \qquad (1-27)$$

where $\langle\phi|$ is a bra vector, and $|\phi'\rangle = a|\phi\rangle$. If the eigenstate corresponding to the minimum eigenvalue W_0 is represented by $|0\rangle$, one has

$$a|0\rangle = 0. \qquad (1-28)$$

Therefore, from Eq.(1-27)

$$\langle 0|\mathcal{H}|0\rangle = 0 + \frac{1}{2}\hbar\omega. \qquad (1-29)$$

This equation represents the result of the proof, i.e. it gives the positive minimum eigenvalue $W_0 = (1/2)\hbar\omega$.

From the second equation of Eq.(1-21), one has

$$\mathcal{H}\, a^{+}|0\rangle = \left(a^{+}\,\mathcal{H} + \hbar\omega a^{+}\right)|0\rangle = \hbar\omega\left(1 + \frac{1}{2}\right)a^{+}|0\rangle. \qquad (1-30)$$

Furthermore, by repeating the same procedure as for Eq.(1-30),

$$\mathcal{H}\left(a^{+}\right)^{n}|0\rangle = \hbar\omega\left(n + \frac{1}{2}\right)\left(a^{+}\right)^{n}|0\rangle. \qquad (1-31)$$

Therefore, $(a^{+})^{n}\,|0\rangle$ corresponds to the energy eigenstate $|n\rangle$ and the corresponding eigenvalue is

$$W_n = \hbar\omega\left(n + \frac{1}{2}\right). \qquad (1-32)$$

It is found from Eqs.(1-16), (1-25) and (1-32) that $a^{+}a$ and $|n\rangle$ are a photon number operator and a photon number eigenstate, respectively, and

$$a^{+}a|n\rangle = n|n\rangle. \qquad (1-33)$$

By noting that $a|n\rangle$ is expressed by using a scalar s_n as in

$$a|n\rangle = s_n|n-1\rangle, \qquad (1-34)$$

the eigenstate $|n\rangle$ given by Eq.(1-25) can be normalized. That is,

$$|s_n|^2\langle n-1|n-1\rangle = \langle n|a^{+}a|n\rangle = n\langle n|n\rangle, \qquad (1-35)$$

from which $s_n = \sqrt{n}$ for the normalization and

$$a|n\rangle = \sqrt{n}|n-1\rangle. \qquad (1-36)$$

By a similar procedure,

$$a^{+}|n\rangle = \sqrt{n+1}\,|n+1\rangle \tag{1-37}$$

Therefore, by using Eq.(1-31), the normalized eigenstate is expressed as

$$|n\rangle = \frac{1}{\sqrt{n!}}\left(a^{+}\right)^{n}|0\rangle \tag{1-38}$$

As a last step on the discussion of the photon number state $|n\rangle$, we consider the coordinate representation of $|n\rangle$, i.e.

$$\phi_n(q) = \langle q|n\rangle. \tag{1-39}$$

Since it is found by using Eqs.(1-14) and (1-28) that

$$(M\omega q + ip)\phi_0(q) = 0 \tag{1-40}$$

one can easily derive by using the momentum operator $p = -i\hbar\, d/dq$ that

$$\frac{\mathrm{d}}{\mathrm{d}q}\phi_0(q) = -\frac{M\omega}{\hbar}q\phi_0(q) \tag{1-41}$$

The solution of this differential equation is expressed as

$$\phi_0(q) = \left(\frac{M\omega}{\pi\hbar}\right)^{\frac{1}{4}} \exp\left[-\frac{1}{2}\left(\frac{M\omega}{\hbar}\right)q^2\right] \tag{1-42}$$

where this function has been already normalized so that the integration value of $|\phi_0(q)|^2$ for $-\infty < q < \infty$ is equal to unity. The expression for $\phi_n(q)$ can be derived by using Eqs.(1-15), (1-38) and (1-42), and is given by

$$\phi_n(q) = \frac{1}{\sqrt{n}}\left(a^{+}\right)^{n}\phi_0(q) = \frac{1}{\sqrt{n!(2M\hbar\omega n!)^{n}}}\left(M\omega q - \hbar\frac{\mathrm{d}}{\mathrm{d}q}\right)^{n}\phi_0(q)$$

$$= \frac{1}{\sqrt{2^{n}n!}}H_n\left(\sqrt{\frac{M\omega}{\hbar}}q\right)\cdot\phi_0(q) \tag{1-43}$$

where H_n represents the Hermite polynomial of the n-th order.

When light stays in the state $|n\rangle$, the expectation value of the electric field operator E_x of Eq.(1-19) is

$$\langle E_x \rangle \equiv \langle n|E_x|n\rangle$$

$$= E_{x0}\sin kz\langle n|a|n\rangle + c\ c = 0 \qquad (1-44)$$

where $c.c.$ represents the complex conjugate. Furthermore, the expectation value of the operator $E_x{}^2$ is

$$\langle E_x^2 \rangle \equiv \langle n|E_x^2|n\rangle$$

$$= E_{x0}^2\sin^2 kz\langle n|\left(a^+a^+ + aa^+ + a^+a + aa\right)|n\rangle$$

$$= 2E_{x0}^2\left(n+\frac{1}{2}\right)\sin^2 kz \qquad (1-45)$$

Equations (1-44) and (1-45) imply that the electric field fluctuates around the zero ensemble average. That is, when the values of electric field are measured for many systems in which the field stays in the state $|n\rangle$, its quantum mechanical average (not the classical time average) of the field amplitude is zero, while its standard deviation

$$\Delta E_x \equiv \sqrt{\langle E_x^2\rangle - \langle E_x\rangle^2}$$

$$= \sqrt{2\left(n+\frac{1}{2}\right)}E_{x0}\sin kz \qquad (1-46)$$

is not zero. It can be interpreted from this result that it is rather difficult to find the wave properties of light through these measurements when light is in the state $|n\rangle$. It should be more reasonable to interpret that there are n pieces of photon of energy $\hbar\omega$, from which it is understood that the eigenstate $|n\rangle$ is called a photon number state. Although a photon is a quanta taking a discrete energy value digitized by the unit $\hbar\omega$, it should be noted that the photon is not spatially and temporally localized but is homogeneously spreaded in the whole volume of the cavity.

If only photons are contained in the cavity, it would not be compatible with the energy conservation principle to decrease or increase the optical energy by a or a^+, which means that annihilation or creation of photons could not take place. To be compatible with the energy conservation principle, it has been implicitly assumed in the discussions described above that a low density atomic vapor is also contained in the cavity. Figure 1-4 shows the two energy levels of atoms relevant to the present discussion. Eigenvalue and eigenstate for these energy levels are represented by W_a, W_b ($W_a > W_b$), $|a\rangle$, and $|b\rangle$, respectively. It is assumed here that one of Bohr's postulates

$$W_a - W_b = \hbar\omega \tag{1-47}$$

holds for the total system composed of the above-mentioned photons and atoms, i.e. that light is exactly resonant to the atoms. If the initial state of the atom and photon is $|b\rangle$ (lower energy) and $|n\rangle$, respectively, the transition of the photon to the state $|n-1\rangle$ by the operator a (see Eq.(1-36)), resulting in the decrease of the photon energy by $\hbar\omega$, means that the atomic energy increases by $\hbar\omega$ (see Eq.(1-47)) owing to the simultaneous transition of the atom to the upper energy state $|a\rangle$ due to the stimulation of the atom by the photon. By these simultaneous decreases and

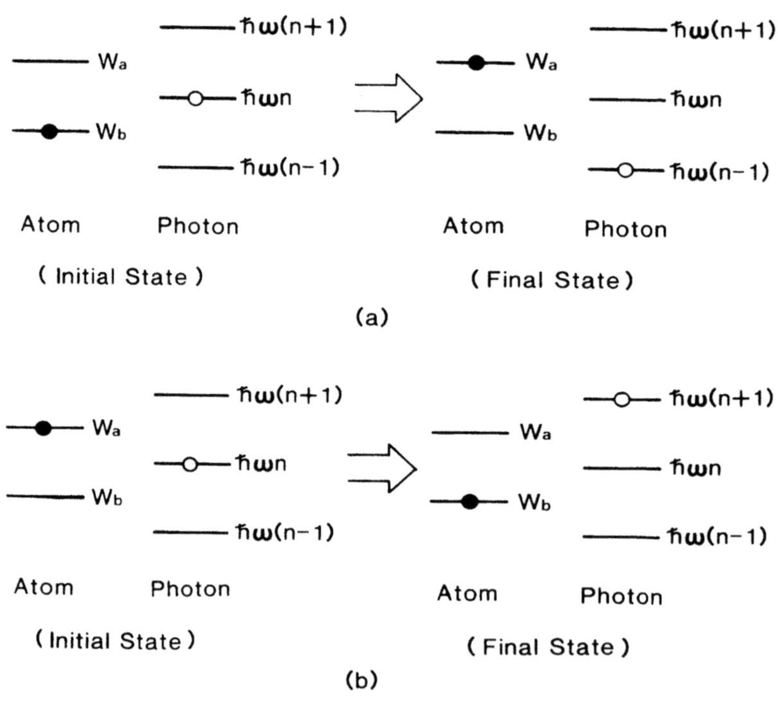

Fig. 1.4. Schematic explanations of the (stimulated) absorption process (a), and stimulated emission process (b), which take place in the system composed of atoms and photons. The closed and open circles represent the atoms and photons, respectively.

increases of energy, energy conservation holds for the total system composed of atoms and photons. These transitions correspond to the (stimulated) absorption of a photon by an atom. On the other hand, the transition of a photon to the state $|n+1\rangle$ by the operator a^{+}, inducing the increase of the photon energy by $\hbar\omega$, means that the atom is de-excited from the upper energy state $|a\rangle$ to the lower energy state $|b\rangle$ and decreases the atomic energy by $\hbar\omega$ as a result of being stimulated by the photon. This process corresponds to the stimulated emission of a photon by the atom, which is opposite to the absorption process described above.

The value $(1/2)\hbar\omega$ in Eqs.(1-29) and (1-32) is called the zero-point energy, which is interpreted as the fluctuation energy of the vacuum field, i.e. corresponding to the process of spontaneously emitting a photon from the upper state atom $|a\rangle$ which is stimulated by the light field fluctuations of Eq.(1-45). This process is called spontaneous emission, i.e., the corresponding to the stimulated emission induced by the vacuum field.

Since light is not always in state $|n\rangle$, its general state can be expressed as

$$|\psi\rangle = \sum_{n=0}^{\infty} c_n |n\rangle \qquad (1-48)$$

One of these states is called the coherent state, $|\alpha\rangle$, if the expansion coefficients c_n have the form

$$c_n \equiv \frac{\alpha^n}{\sqrt{n!}} \exp\left(-|\alpha|^2/2\right) \qquad (1-49)$$

where α is a complex number. The quantity $|c_n|^2$ gives the probability of finding that the light is in the state $|n\rangle$, which is expressed as

$$|c_n|^2 = \frac{|\alpha|^{2n}}{n!} \exp\left(-|\alpha|^2\right) \qquad (1-50)$$

This expression has been well known in probability theory because it is identical to the Poisson distribution. By substituting Eqs.(1-38) and (1-49) into Eq.(1-48), the coherent state is defined by the expression

$$|\alpha\rangle = e^{-\frac{|\alpha^2|}{2}} \sum_{n=0}^{\infty} \frac{\alpha^2}{\sqrt{n!}} \left(a^{+}\right)^n |0\rangle \qquad (1-51)$$

It can be easily found that the expectation value of the photon number $\langle n\rangle$ for this state is

$$\langle n \rangle = \langle \alpha | a^+ a | \alpha \rangle = |\alpha|^2 \tag{1-52}$$

For the discussions described below, it should be noted that the following relation holds

$$\left[a \ \exp\left(\alpha a^+\right)\right] = \alpha \exp\left(\alpha a^+\right) \tag{1-53}$$

where $\exp(\alpha a^+)$ is a polynomial composed by the operators, which is defined by

$$\exp\left(\alpha a^+\right) \equiv \sum_{n=0}^{\infty} \frac{1}{n!}\left(\alpha a^+\right)^n \tag{1-54}$$

The validity of Eq.(1-53) can easily be confirmed by using Eq.(1-17). After transforming Eq.(1-53) into

$$a \ \exp\left(\alpha a^+\right) = \alpha \exp\left(\alpha a^+\right) + \exp\left(\alpha a^+\right) a \tag{1-55}$$

applying both sides of this relation on $|0\rangle$ and using Eq.(1-28), one can derive the relation

$$a|\alpha\rangle = \alpha|\alpha\rangle \tag{1-56}$$

This relation implies that the coherent state is an eigenstate of the operator a and its eigenvalue is α. Furthermore, the following relations hold for the coherent state.

$$\langle \alpha | a^+ = \alpha * \langle \alpha |$$

$$\langle \alpha | a | \alpha \rangle = \alpha \tag{1-57}$$

$$\langle \alpha | a a^+ | \alpha \rangle = |\alpha|^2$$

where * represents the complex conjugate, as was the case of *c.c.* Calculations similar to Eqs.(1-44) and (1-45) can easily give

$$\langle E_x \rangle \equiv \langle \alpha | E_x | \alpha \rangle$$

$$= E_{x0}(\alpha + \alpha *)\sin kz$$

$$(1-74)$$

$$\left\langle E_x^2 \right\rangle \equiv \langle \alpha | E_x^2 | \alpha \rangle$$

$$= E_{x0}^2 \left\{ (\alpha + \alpha *)^2 + 1 \right\} \sin^2 kz$$

These relations mean that the quantum mechanical expectation value of E_x takes the non-zero value, i.e. the wave properties of light can be observed by measurements. However, its standard deviation takes a non-zero value of $\Delta E_x = E_{x0} \sin kz$, which is due to the spontaneous emission fluctuations.
One may show easily that the cross-correlation coefficient

$$\gamma \equiv \frac{\langle \psi | a | \psi \rangle \langle \psi | a^+ | \psi \rangle}{\langle \psi | a^+ a | \psi \rangle} \qquad (1-59)$$

is not larger than unity for a general state vector $|\psi\rangle$. However, $\gamma = 1$ for the coherent state $|\alpha\rangle$, meaning that, as was previously discussed by using Eqs.(1-23) and (1-24), the two running-waves propagating towards $+z$ and $-z$ directions are strongly correlated with each other. In the other words, an interference fringe or a beat signal between the two waves can be observed, which corresponds to the conclusion of the simple discussions in 1.1.

Further properties of the coherent state can be derived. If an observable physical quantity with an operator X is measured when light is in the state $|\psi\rangle$, a measure of uncertainty in the measurement is the standard deviation (see Eq.(1-46)), which is expressed as

$$\Delta X_j \equiv \sqrt{\left\langle X_j^2 \right\rangle - \left\langle X_j \right\rangle^2} \qquad (j = 1\ 2) \qquad (1-60)$$

where

$$\left\langle X_j^2 \right\rangle = \langle \psi | X_j^2 | \psi \rangle$$

$$\left\langle X_j \right\rangle = \langle \psi | X_j | \psi \rangle$$

The Hermitian operators X_1 and X_2, which do not commute, cannot be measured simultaneously with absolute exactitude. As a result, there is always some uncertainty left. The relationship between the uncertainties ΔX_1 and ΔX_2 can easily be derived as

$$\Delta X_1 \Delta X_2 \geq \frac{1}{2}\left|\langle [X_1 \ X_2] \rangle\right|$$

$$\equiv \frac{1}{2}\left|\langle \psi | [X_1 \ X_2] | \psi \rangle\right| \qquad (1-61)$$

This is a well known Heisenberg uncertainty relationship. For example, the substitution of Eq.(1-13) into this equation gives

$$\Delta p \Delta q \geq \frac{\hbar}{2} \qquad (1-62)$$

for the position and momentum operators q and p.

The expression of the magnetic field vector can be derived by using Eqs.(1-9), (1-12), (1-14) and (1-15), which is expressed as

$$H_y(z \ t) = 2i\sqrt{\frac{\varepsilon_0}{\mu_0}} E_{x0}\left(a - a^+\right)\cos kz \qquad (1-63)$$

This magnetic field and the electric field given by Eq.(1-19) are Hermitian operators because the operators $(a + a^+)$ and $(a - a^+)$ are Hermitians. Furthermore, by noting the relationship derived by Eq.(1-17)

$$\left[\left(a + a^+\right) \ \left(a - a^+\right)\right] = -2 \qquad (1-64)$$

Eq.(1-61) gives

$$\Delta\left(a + a^+\right)\Delta\left(a - a^+\right) \geq 1 \qquad (1-65)$$

Especially, for the coherent state,

$$\Delta\left(a + a^+\right)\Delta\left(a - a^+\right) = 1 \qquad (1-66)$$

which means that the coherent state is the minimum uncertainty state satisfying Eq.(1-66). It also means that the relationship between the uncertainties of the electric and magnetic fields, derived from Eqs.(1-19), (1-63) and (1-65),

$$\Delta E_x \Delta H_y \geq \frac{\hbar}{2}\left(\frac{2\omega^2}{kV}\right) \qquad (1-67)$$

takes the minimum. Since the classical status of E_x and H_y are the waves with a $\pi/2$ phase difference, the minimum uncertainty of Eq.(1-67) corresponds to the classical status of the wave with the minimum fluctuations.

When the photon number is large enough, i.e. $\langle n \rangle >> 1$, a commutation relationship between the photon number n and the phase operators ϕ holds, from which the uncertainty relationship can be derived as

$$\Delta n \Delta \phi \geq 1/2 \qquad (1-68)$$

For this relationship, the coherent state also takes the minimum uncertainty. Especially, since the coherent state gives

$$\Delta n = \sqrt{\left\langle \left(a^+ a\right)^2 \right\rangle - \left\langle a^+ a \right\rangle^2}$$

$$= |\alpha| = \sqrt{\langle n \rangle}\left(\equiv \sqrt{\langle \alpha | a^+ a | \alpha \rangle}\right) \qquad (1-69)$$

it leads to

$$\Delta n / \langle n \rangle = 1/\sqrt{\langle n \rangle} \qquad (1-70)$$

and from Eq.(1-68)

$$\Delta \phi = 1/2\sqrt{\langle n \rangle} \qquad (1-71)$$

This relationship means that precise and simultaneous measurements of the amplitude (or power) and phase of the light are possible for the coherent state within the limit of $\langle n \rangle >> 1$.

On the other hand, since the physical implications of Eq.(1-68) are lost for $\langle n \rangle < 1$, a more general uncertainty relationship, valid for $\langle n \rangle < 1$, has been derived [1] by using the alternative operators S and C, which are defined by

$$S \equiv \frac{1}{2i\sqrt{a^+ a + 1}}\left(a - a^+\right) \qquad (1-72)$$

and

$$C \equiv \frac{1}{2i\sqrt{a^{+}a+1}}\left(a+a^{+}\right) \qquad (1-73)$$

Since the operators S and C are Hermitian and are related to the phase ϕ of a classical lightwave in such a manner as

$$S = \sin\phi$$
$$C = \cos\phi \qquad (1-74)$$

within the limit of $\langle n \rangle >> 1$, they can be interpreted as operators representing a phase of the optical field. The uncertainty relationship between S, C and the photon number, is given by

$$(\Delta n)^2 \frac{(\Delta S)^2 + (\Delta C)^2}{\langle S \rangle^2 + \langle C \rangle^2} \geq \frac{1}{4} \qquad (1-75)$$

which is more accurate than that of Eq.(1-68). Furthermore, the following uncertainty relationship holds for ΔS and ΔC

$$\Delta S \Delta C \geq (1/4)\exp(-\langle n \rangle) \qquad (1-76)$$

The solution of a time-independent Schrödinger equation can be derived if the coherent state of Eq.(1-51) is expressed by the coordinate representation, as was the procedure given by Eq.(1-39). The solution of a time-dependent Schrödinger equation can also be derived by replacing $|\alpha\rangle$ by $|\alpha\rangle \exp(-i\omega t)$ in the solution of the time-independent Schrödinger equation. This time-dependent solution is given by [B]

$$\phi_{\alpha}\left(E_x\ t\right) = \frac{1}{\sqrt{\sqrt{2\pi}E_{x0}}} \exp\left[\frac{1}{2}\left(\alpha e^{-i\omega t}\right)^2 - \frac{1}{2}|\alpha|^2\right]$$

$$\exp\left\{-\frac{1}{2}\left[\frac{1}{\sqrt{2}E_{x0}}E_x - \sqrt{2}\alpha e^{-i\omega t}\right]^2\right\} \qquad (1-77)$$

where q has been replaced by E_x by using Eq.(1-8). Therefore, the probability density finding that the value of the electric field amplitude is E_x at the time t is

$$|\phi_\alpha(E_x\ t)|^2 = \frac{1}{\sqrt{2\pi E_{x0}}}\exp\left\{-\left[\frac{1}{\sqrt{2E_{x0}}}E_x - \sqrt{2}\alpha\cos(\omega t+\theta)\right]^2\right\} \tag{1-78}$$

where

$$\alpha = |\alpha|\exp(-i\theta) \tag{1-79}$$

Equation (1-78) represents a Gaussian wave packet with the center of $\sqrt{2}E_{x0}|\alpha|\cos(\omega t+\theta)$ and the half width at half-maximum of $\sqrt{2}E_{x0}$. That is, as is shown by Fig. 1-5(a)*, the packet bounces back and forth sinusoidally without changing its width. In other words, the coherent state maintains an uncertainty minimum and its packet of probability density "coheres". By these meanings, the coherent state resembles the classical field as closely as quantum mechanics permits. For references, it is shown by Fig. 1-5(b)* that a packet of general state of light varies in time.

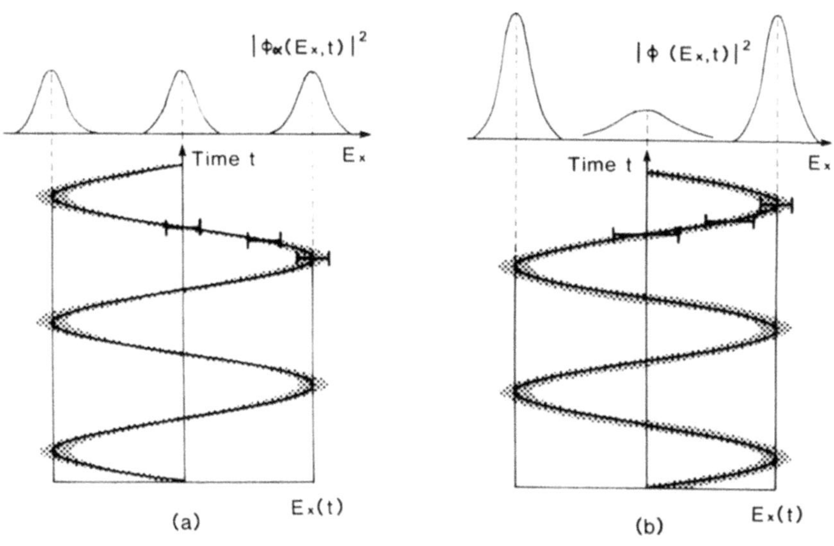

Fig. 1.5. Probability density of the electric field amplitude of a light (upper), and temporal variations of the electric field with fluctuations (lower) [B]. (a) Coherent state. (b) General state.

*M. Sargent III, M. C. Scully, and W. E. Lamb, Jr., *Laser Physics*, 1974 by Addison-Wesley Publishing Company, Inc. Reprinted with permission of the publisher.

[Textbooks]

A) K. Shimoda, T. Yajima, Y. Ueda, T. Shimizu and T. Kasuya, *Quantum Electronics,* Shokabo Publishing, Tokyo, 1972 (in Japanese).
B) M. Sargent, III, M. O. Scully and W. E. Lamb, Jr., *Laser Physics,* Addison-Wesley, Reading, 1974.
C) H. Haken, Light, Vol. 1, *Waves, Photons, Atoms,* North-Holland, Amsterdam, 1981.
D) R. Loudon, *The Quantum Theory of Light,* Clarendon Press, Oxford, 1973.
E) K. Shimoda, *Introduction to Laser Physics*, 2nd ed., Springer-Verlag, Berlin, 1986.

[References]

1) L. Susskind and J. Glowgower, "Quantum Mechanical Phase and Time Operator", *Physics.*, Vol. 1 (1964) pp. 49–62.

Chapter 2

Principles of Laser Oscillation

Nearly coherent light can be generated from a laser. Principles of laser oscillation and their theoretical formulation are described in this chapter. In laser oscillations, the spontaneously emitted light from atoms is used as a trigger and it is subsequently amplified by inducing the stimulated emission from other atoms. This amplified light is positively fedback to the cavity, in which the atoms and light are confined, shown in Fig. 2-1(a) [D]; as a result, a stable self-sustained oscillation is achieved.

2.1 Cavity

The cavity used for the positive feedback of light is a type of resonator whose inner walls are made of high reflective plate, as shown in Fig. 2-1(a). From the principle of electro-magnetic field theory, the tangential component of an electric field vector must be zero on the wall surface in order that this field may be confined stationarily in the cavity. By this boundary condition, the spatial profile and the value of frequency or the wavelength of this field in the cavity are determined. Figures 2-1(c) and (d) show several spatial profiles satisfying this condition along the x- and y-axes in the cavity. By using the integers m and n which describe the numbers of intersections of this profile with the x- and y-axes, these fields are called the mode with orders of 0th, 1st, 2nd, ... and so on, i.e. there exists a "transverse mode of the (m, n) order" in the cavity. The word "transverse" can be understood from Fig. 2-1(a), i.e. the x-y plane corresponds to the transverse of the longitudinal direction of the cavity (the z axis) along which the laser beam is extracted as an output. The boundary conditions described above implies that a standing wave is formed as a result of the interference between the two running waves propagating toward the $+x$ and $-x$ axes. It also holds along the y-axis. Figure 2-2(a) shows several transverse mode patterns observed by a gas laser [1].

Since the optical wavelength is much shorter than the cavity dimension, there is a great number of transverse modes for the cavity configuration of Fig. 2-1(a), which is sometimes inconvenient for practical uses. To avoid this inconvenience, a modified cavity in which the side walls shown in Fig. 2-1(a) have been removed

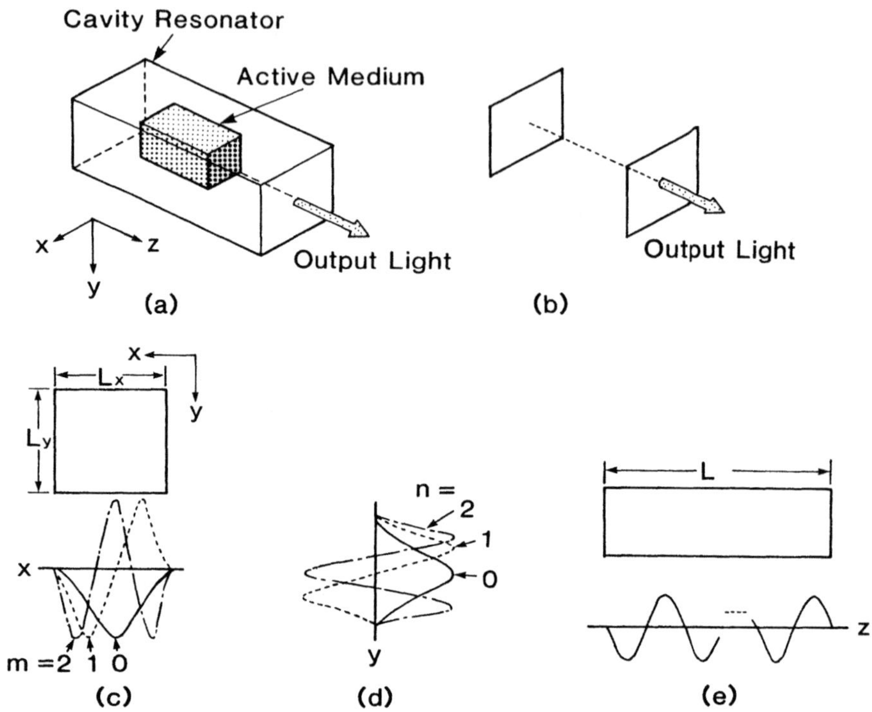

Fig. 2.1. Configuration and characteristics of a cavity [D]. (a) Basic configuration. (b) Fabry-Perot cavity. (c) Electric field distribution along the x-axis in cavity. (d) Electric field distribution along the y-axis in cavity. (e) Electric field distribution along the z-axis in cavity.

is used, which is called a Fabry-Perot resonator, as is shown in Fig. 2-1(b). By this configuration, only the light propagating along the z-axis is confined efficiently in the cavity, by which the number of transverse modes is drastically reduced. However, if the optical confinement of the transverse directions is insufficient, the light propagating along the z-axis leaks out from the cavity due to the diffraction effect. To avoid this leakage, concave mirrors are usually used in the Fabry-Perot cavity. The condition of stable optical confinement provided by these concave mirrors, or the condition of stable cavity, is given by

$$0 \le \left(1 - L / r_1\right)\left(1 - L / r_2\right) \le 1 \qquad (2-1)$$

where r_1 and r_2 are the radius of curvature of each concave mirror, respectively, and L is the cavity length along the z-axis. For a stable Fabry-Perot cavity composed of two square mirrors with the same radius of curvature, the spatial profile of the optical electric field amplitude shown in Figs. 2-1(c) and (d) is expressed as

$$E(x\ y\ z) \propto H_m(bx)H_n(by)$$

$$\exp\left\{-\left(x^2+y^2\right)\left(\frac{1}{w^2}+\frac{ik}{2R}\right)-ikz\right\} \qquad (2-2)$$

where H_m and H_n are the Hermite polynomials of the m-th and n-th orders, re-

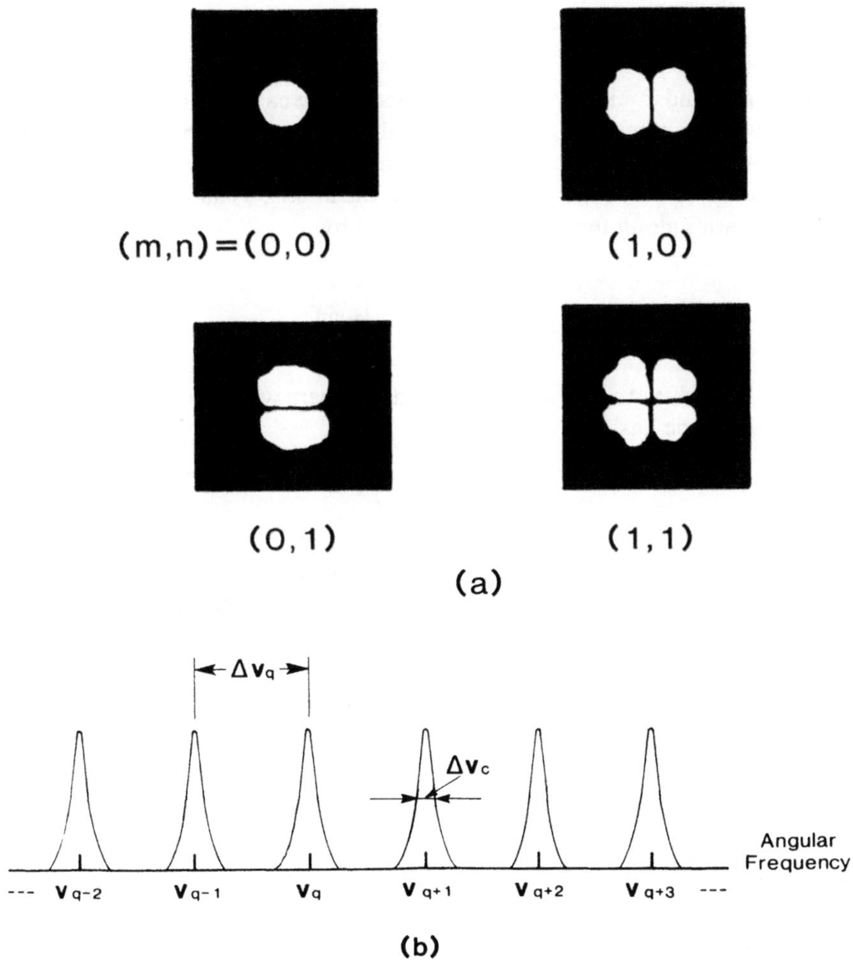

Fig. 2.2. Transverse and longitudinal modes [D]. (a) Cross-sectional profile of transverse modes(after Rigrod [1]). (b) Spectral profiles of longitudinal modes.

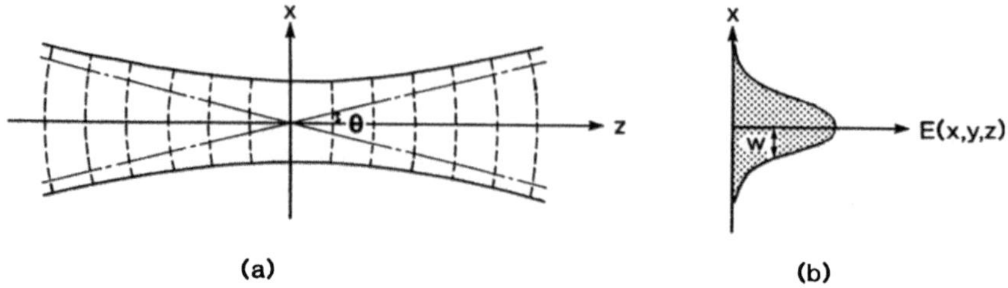

Fig. 2.3. Contour of the electric field of the (0, 0)-order transverse mode. (a) θ: Divergence angle (see Eq.(2-5)). Wavefront is shown by broken curves (see Eq.(2-6)). (b) Cross-sectional profile. w: The spot size (see Eq.(2-3)).

spectively, b, w and R take a value determined by the cavity configuration, and k is the optical wave number. The quantity w is called a spot size, representing the radius of the cross-section in the x-y plane at which the value of $E(x, y, z)$ is down by a factor e^{-1} compared to its value at $x = y = 0$, i.e. on the z-axis, as shown in Fig. 2-3. The spot size w varies along the z-axis and is given by

$$w = w_0\sqrt{1+\left(2z / kw_0^2\right)^2} \qquad (2-3)$$

where w_0 is the value at $z = 0$, which is called the beam waist. For a confocal cavity, i.e., $r_1 = r_2 = L$, one has

$$w_0 = \sqrt{L / k} \qquad (2-4)$$

The transverse mode profile observed far from the cavity end is called the far field pattern and the divergence angle of the optical beam in the far field region is given by

$$\theta \equiv \lim_{z\to\infty} \frac{w}{z} = \frac{2}{kw_0} \qquad (2-5)$$

The quantity R in Eq.(2-2) represents the radius of curvature of a constant phase front, i.e. in the paraxial approximation along the z-axis, the constant phase front can be approximated as a spherical surface with its center at z - R and with a radius of $|R|$. Since R is given by

$$R = z\left\{1+\left(kw_0^2 / 2z\right)^2\right\} \qquad (2-6)$$

the plane wave is realized at $z = 0$ and ∞.

The spatial profile of the electric field along the z-axis, shown in Fig. 2-1(e), is also determined by the same boundary condition as those for the x- and y-axes. The number of intersection, q, of this profile with the z-axis is quite large because the cavity dimension along the z-axis is generally much larger than the optical wavelength. One can easily obtain this number q by noting that a standing wave is formed along the z-axis, and is expressed as

$$\left.\begin{aligned} \frac{\lambda}{2n_q}q &= L \\ \nu_q &= \frac{2\pi c}{\lambda_q} \end{aligned}\right\} \qquad (2-7)$$

where λ_q and ν_q are the resonance wavelength and angular frequency of the standing wave in the cavity, respectively, n_q is the refractive index of the medium in the cavity for this standing wave and c is the speed of light in vacuum. Given by this equation, the light fields with the wavelength λ_q and the angular frequency ν_q are allowed to exist stationarily in the cavity. Although the values of the wavelengths and angular frequencies depend on the transverse mode numbers m and n, their dependence is usually less than that on q because the optical confinement effects along the x- and y-axes are weaker than those along the z-axis in the Fabry-Perot cavity. The electric field identified by the number q is called the longitudinal mode of the q-th order. Therefore, by combining it with the previously discussed transverse modes, it can be said that there exist modes of the (m, n, q)-th order in the cavity. It means that among the initially existing lights in the cavity, only those lights satisfying the boundary conditions mentioned above can survive after running back and forth along the x-, y- and z-axes and are allowed to exist as a stable resonant field in the form of a mode of the (m, n, q)-th order.

The separations of the resonant wavelength and angular frequency between adjacent longitudinal modes are easily obtained from Eq.(2-7) under the approximation of $q >> 1$, which is given by

$$\Delta\lambda_q = \frac{\lambda_q^2}{2n_q L}$$

$$\Delta\nu_q = \frac{\pi c}{n_q L} \qquad (2-8)$$

This relationship shows that the resonant angular frequency of each mode is distributed on the frequency axis with an equal separation of $\pi c/n_q L$ as is shown in Fig. 2-2(b).

Figure 2-2(b) shows that each spectral profile representing the longitudinal mode has an angular frequency width of $\Delta\nu_c$. It can easily be shown that this value depends on the magnitude of the cavity loss. When the light propagates toward the z-axis, its energy density per unit volume $W_q(z)$ decreases due to the cavity loss, which can be expressed as

$$W_q(z) = W_q(0)\exp(-\alpha_{tl}z). \tag{2-9}$$

The quantity α_{tl} is the cavity loss per unit length given by

$$\alpha_{tl} = \alpha_l + \frac{1}{2L}\ln\left(\frac{1}{R_1 R_2}\right), \tag{2-10}$$

where α_l is the internal loss determined by the absorption, and/or scattering by the medium in the cavity, the leakage by diffraction effect and so on, R_1 and R_2 are the power reflectivities of the two mirrors. Time dependence of the energy decrease at one position on the z-axis is derived from Eq.(2-9) and is expressed as

$$W_q(t) = W_q(0)\exp(-\alpha'_{tl}\, t), \tag{2-11}$$

where α_{tl}' represents the cavity loss per unit time given by

$$\alpha'_{tl} = \alpha_{tl} c / n_q \tag{2-12}$$

From Eq.(2-11), one has the expression for the temporal decrease of the optical field amplitude with an angular frequency of Eq.(2-7), which is expressed as

$$E_x(t) = \sqrt{2W_q(0)/\varepsilon_0}\exp\{-(\alpha'_{tl}/2)t + i\nu_q t\}. \tag{2-13}$$

The Fourier transform of this relationship gives the spectral profile of the optical field existing in the cavity, which corresponds to the spectral profile of the longitudinal mode (resonance spectral shape of the cavity). It is expressed as

$$I_c(\nu) \propto \frac{1}{(\nu-\nu_q)^2 + (\alpha'_{tl}/2)^2}. \tag{2-14}$$

In this expression, the full width at the half maximum (FWHM) of the profile $\Delta\nu_c$

is given by

$$\Delta\nu_c = \alpha'_{tl} = \alpha_{tl}c / n_q \,, \tag{2-15}$$

which is proportional to the cavity loss.

Equation (2-14) means that the optical field whose angular frequency is deviated from ν_q within the range of $\pm\Delta\nu_c/2$ can exist in the cavity even though the confinement of the cavity for this field is lower than that for the exactly resonant field with an angular frequency of ν_q.

By using Eq.(2-15), one can define the quality factor Q_q of the cavity, which is expressed as

$$Q_q \equiv \nu_q / \Delta\nu_c = n_q \nu_q / \alpha_{tl}c. \tag{2-16}$$

On the other hand, the average time during which light is confined in the cavity, or the time during which its light power is down by a factor e^{-1} as compared to its initial power value, is found from Eq.(2-1) and given by

$$\tau_p = 1/\alpha'_{tl} = \left(n_q / c\right) / \left\{\alpha_l + \frac{1}{2L}\ln\left(\frac{1}{R_1 R_2}\right)\right\}, \tag{2-17}$$

which is called the photon lifetime.

2.2 Light Amplification and Laser Oscillation

To estimate the magnitude of spontaneous emission, stimulated emission and absorption by atoms confined in the cavity, interactions between the electrons in atoms and the photons are discussed by electric dipole interaction approximation. For the purpose of this discussion, a semiclassical model is employed [A, B], which is different from that employed in 1.2, i.e., the atoms are described by quantum mechanics but light is treated as a wave described by Maxwell's equations.

The electric field vector of the light which is incident upon the atom is expressed as

$$E_x(t) = E_{x0}\cos\nu t. \tag{2-18}$$

Here we assumed that the electric field is linearly polarized along the x-axis, as in 1.2. Although the angular frequency ν should be nearly equal to the Bohr angular frequency ω of Eq.(1-46), it is not necessary for it to be exactly equal to ω. The perturbation energy $\mathcal{V}$ due to the electric-dipole interaction is therefore given by

$$\mathcal{V} = (-ex) \cdot E_x, \tag{2-19}$$

where e is the electron charge and x is the x-axis component of the position vector of the electron with respect to the nucleus. The Schrödinger's time-dependent equation for an electron subject to such a perturbation is

$$i\hbar \frac{\mathrm{d}}{\mathrm{d}t} | \psi \rangle = (\mathcal{H}_0 + \mathcal{V}) | \psi \rangle, \tag{2-20}$$

where $|\psi\rangle$ is the state vector of the electron and $\mathcal{H}_0$ is the unperturbed Hamiltonian of the electron following the eigenvalue equation

$$\mathcal{H}_0 | a \rangle = W_a | a \rangle, \qquad \mathcal{H}_0 | b \rangle = W_b | b \rangle. \tag{2-21}$$

The state vector of the electron under the perturbation in Eq.(2-20) can be expressed as

$$| \psi \rangle = c_a | a \rangle \cdot \exp\{-i(W_a / \hbar)t\}$$

$$+ c_b | b \rangle \cdot \exp\{-i(W_b / \hbar)t\}. \tag{2-22}$$

Here, it is assumed that the eigenvectors $|a\rangle$ and $|b\rangle$ have an opposite parity and are orthonormal, i.e.

$$\langle x | y \rangle = \delta_{xy} \qquad (x, y = a, b), \tag{2-23}$$

where δ_{xy} is Kronecker's delta. Furthermore, the vector $|\psi\rangle$ is also assumed to be normalized, i.e.

$$|c_a|^2 + |c_b|^2 = 1. \tag{2-24}$$

By substituting Eq.(2-22) into Eq.(2-20), and using Eqs.(2-21), (2-23) and (2-24), one obtains

$$\frac{\mathrm{d}c_a}{\mathrm{d}t} = \frac{ipE_{x0}}{2\hbar} \left\{ e^{i(\omega - \nu)t} + e^{i(\omega + \nu)t} \right\} c_b$$

$$\frac{\mathrm{d}c_b}{\mathrm{d}t} = \frac{ipE_{x0}}{2\hbar} \left\{ e^{-i(\omega - \nu)t} + e^{-i(\omega + \nu)t} \right\} c_a, \tag{2-25}$$

where p is the magnitude of the electric-dipole matrix element defined by

$$p \equiv e\langle a|x|b\rangle. \qquad (2-26)$$

Since the second term on the right hand side of Eq.(2-25) varies much faster than the first term because $(\omega + \nu) >> (\omega - \nu)$, the second term can be neglected when Eq.(2-5) is integrated with respect to time. This is called the rotating-wave approximation.

By using this approximation and imposing the initial condition of $c_a(0) = 1$ and $c_b(0) = 0$ (i.e., the electron stays in the upper energy level at $t = 0$), the value of $c_b(t)$ is obtained from Eq.(2-25), which is expressed as

$$c_b(t) = \frac{ip}{\hbar\mu} E_{x0} e^{-i(\omega-\nu)t/2} \sin(\mu t / 2), \qquad (2-27)$$

where μ is called the Rabi flopping frequency defined by

$$\mu \equiv \sqrt{(\omega - \nu)^2 + (pE_{x0} / \hbar)^2}. \qquad (2-28)$$

The probability of finding an electron in the lower state at time t is obtained from Eq.(2-27) and is given by

$$|c_b|^2 = \frac{\sin^2(\mu t / 2)}{(\mu t / 2)^2} \left(\frac{pE_{x0}}{\hbar}\right)^2. \qquad (2-29)$$

It represents the probability of de-excitation of the electron from the upper to the lower state, i.e. the probability that the stimulated emission of light be proportional to $(p/\hbar)^2$. By further detailed discussions, the measure representing the magnitude of the stimulated emission is defined by

$$B \equiv \frac{\pi p^2}{6\hbar^2 \varepsilon_0}, \qquad (2-30)$$

which is called the Einstein B coefficient which is proportional to $(p/\hbar)^2$, as was the case of $|c_b|^2$ in Eq.(2-29).

By substituting $c_b(t)$ of Eq.(2-27) and the similarly derived $c_a(t)$ into Eq.(2-22), the expectation value of the dipole moment of the atom induced by light can be obtained and is expressed as

$$\langle ex\rangle \equiv \langle \psi|ex|\psi\rangle = p^* c_a c_b^* \exp(-i\omega t) + c.c.$$

$$= \frac{p^2 E_{x0}}{\hbar\mu^2}\left[-(\nu-\omega)(1-\cos\mu t)\cos\upsilon t - \mu\sin\mu t\sin\nu t\right]. \tag{2-31}$$

By noting that $\mu << \nu$ in most cases (i.e., $\nu \cong \omega$ and $pE_{x0}/\hbar << 1$), it can be easily seen from this equation that the dipole moment oscillates with the same angular frequency ν as that of the incident light because the principal term of this equation is sin νt. Therefore, the secondary light emitted from this oscillating dipole could have the same angular frequency ν as that of the incident light.

Then, it is easily confirmed that the dipole moment of Eq.(2-31) emits the secondary light. By assuming $\mu << \nu$ again, Eq.(2-31) becomes

$$\langle ex\rangle \propto -\sin\nu t = \cos\left(\nu t - \frac{\pi}{2}\right), \tag{2-32}$$

which implies that the phase of the dipole moment shows a delay of $\pi/2$ radian when compared to that of incident light. In this case, energy transferred from light to atom, i.e.

$$W = \frac{1}{(1/\nu)}\int_0^{1/\nu} \frac{\mathrm{d}}{\mathrm{d}t}\langle ex\rangle E_x \mathrm{d}t$$

$$\propto -\nu\int_0^{1/\nu} \cos^2 \nu t \mathrm{d}t = -\nu/2, \tag{2-33}$$

takes a negative value; consequently, this means that light emission takes place. By the discussions given above, it was demonstrated that light emitted from an atom has the same angular frequency as that from incident light, which corresponds to the semiclassical expression of stimulated emission. Moreover, the emission of secondary light with an identical angular frequency as that of incident light implies an amplification of incident light. By further detailed discussions [D], it is also found that the direction of propagation of this secondary light is the same as that of incident light. Therefore, it can be confirmed that light amplification is possible by using the stimulated emission process.

One can easily find that the $|c_a|^2$, i.e. the probability of finding an electron in the upper level, or in other words, the absorption probability, takes on the same value as that given by Eq.(2-29) by solving Eq.(2-25) under the initial condition of $c_a(0) = 0$ and $c_b(0) = 1$. This means that the absorption probability is equal to the stimulated emission probability as has been described in 1.2. However, in that case it is found that the expectation value of the dipole moment has a phase lead of $\pi/2$ radian when compared to that of incident light and that the value of W in Eq.(2-33) is positive, which represents the fact that light is really absorbed by the atom.

In addition to stimulated emission and absorption, spontaneous emission should be taken into account when this system maintains its thermal equilibrium in a steady state. The spontaneous emission probability can be estimated by considering a system composed of atoms and photons which interact mutually in a closed vessel, or a cavity resonator. For this estimation, a set of equations is used to describe temporal variations of the atom populations in the upper and lower energy levels, i.e. n_a and n_b, respectively. These equations are called a set of rate equations and are given by

$$\mathrm{d}n_a / \mathrm{d}t = -BU(\omega)n_a + BU(\omega)n_b - An_a$$

$$\mathrm{d}n_b / \mathrm{d}t = BU(\omega)n_b - BU(\omega)n_b + An_a \;, \qquad (2-34)$$

where the first, second and third terms on the right-hand sides represent the changing rate of atom populations due to stimulated emission, absorption, and spontaneous emission, respectively. The quantity $U(\omega)$ is the energy density of light per unit volume and unit angular frequency interval, which corresponds to the spectral distribution of blackbody radiation. That is, the energy density per a unit volume in the cavity within an angular frequency interval $\omega \sim \omega + \mathrm{d}\omega$ is given by the product of the number of possible modes in the cavity

$$m(\omega)\mathrm{d}\omega = \frac{\omega^2}{\pi c^3}\mathrm{d}\omega, \qquad (2-35)$$

and the mode energy of each mode

$$W(\omega) = \sum_{n=0}^{\infty} n\hbar\omega e^{-n\hbar\omega/k_B T} \Big/ \sum_{n=0}^{\infty} e^{-n\hbar\omega/k_B T}, \qquad (2-36)$$

where k_B is Boltzmann's constant and T is the absolute temperature. The final expression is

$$U(\omega)\mathrm{d}\omega = m(\omega)\mathrm{d}\omega W(\omega)$$

$$= \frac{\omega^2}{\pi c^3} \cdot \frac{\hbar\omega}{e^{\hbar\omega/k_B T} - 1}\mathrm{d}\omega. \qquad (2-37)$$

This expression is popularly known as Planck's radiation law.

In a steady state, $\mathrm{d}/\mathrm{d}t = 0$ in Eq.(2-34); furthermore, at thermal equilibrium, populations are related by Boltzmann's distribution

$$n_a / n_b = \exp(-\hbar\omega / k_B T) \qquad (2-38)$$

By substituting this equation into Eq.(2-34), one has

$$U(\omega) = (A / B) / \left(e^{\hbar\omega / k_B T} - 1\right) \qquad (2-39)$$

By comparing this equation with Eq.(2-37) and by using Eq.(2-30), one has

$$A = \frac{p^2 \omega^3}{6\pi\hbar\varepsilon_0 c^3} \qquad (2-40)$$

This quantity represents the magnitude of the spontaneous emission probability and is called the Einstein A coefficient. This equation shows that the value of A increases in proportion to ω^3, which means that contribution from the spontaneous emission is more important in the optical frequency range than in the microwave frequency range and that it is not easy to develop a very short wavelength laser, e.g. an X-ray laser.

The ensemble of atoms plays only the role of an absorbing medium at thermal equilibrium because n_b is larger than n_a in Eq.(2-38) and the probabilities of stimulated emission and absorption are equal, as already known from Eqs.(2-29)–(2-33). In order to use this ensemble as an amplifying medium, the population inversion, i.e. $n_a > n_b$, should be realized by exciting some electrons from the lower state to the upper state through an external power source. If the expression of Eq.(2-38) is still used in such a situation of population inversion, the equivalent temperature T takes on a negative value due to which the population inversion is sometimes called a negative temperature condition. In this condition, the ensemble of atoms plays the role of an amplifying medium and laser oscillation can be realized if this medium is confined in the low loss optical cavity. However, the laser apparatus is not a closed system any longer since it is connected to an external power source and also the external field in a degree determined by the cavity loss. That is, the laser is a non-equilibrium and open system which interacts with the reservoir.

Immediately after the power source has been turned on, atoms are stimulated by vacuum fluctuation and as a result, emit light spontaneously. This spontaneously emitted light induces stimulated emission from other atoms and light generated by this stimulated emission has the same angular frequency and propagates to in the same direction as that of incident light, i.e. initially spontaneously emitted light. This emitted light induces further stimulated emission to other atoms and by repeating this process, the initially emitted light power is amplified.

The gain in light amplification is determined by the magnitude of the total stimulated emission from the ensemble of atoms. As shown by Fig. 2-4(a), the center of the gain spectral profile corresponds to the Bohr angular frequency ω of Eq.(1-47) [D]. Its width $\Delta\nu_G$ depends on the lifetimes of the electronic transition, induced dipole moments, and so on. Furthermore, its peak height is determined by the energy supplied from the external power source, i.e. by the value of population inversion. If the value of the peak gain at the center frequency of its spectral profile is larger than that of the cavity loss in Eq.(2-10), a net amplification of light is realized and a stationary oscillation takes place due to the positive feedback by the cavity. The minimum gain G_{th} required for this laser oscillation is therefore given by

$$G_{\rm th} = \alpha_{\rm tl} \qquad (2-41)$$

which is called the threshold gain.

Since the light emitting source described above utilizes "Light Amplification by Stimulated Emission of Radiation", it is called a LASER. The laser oscillation

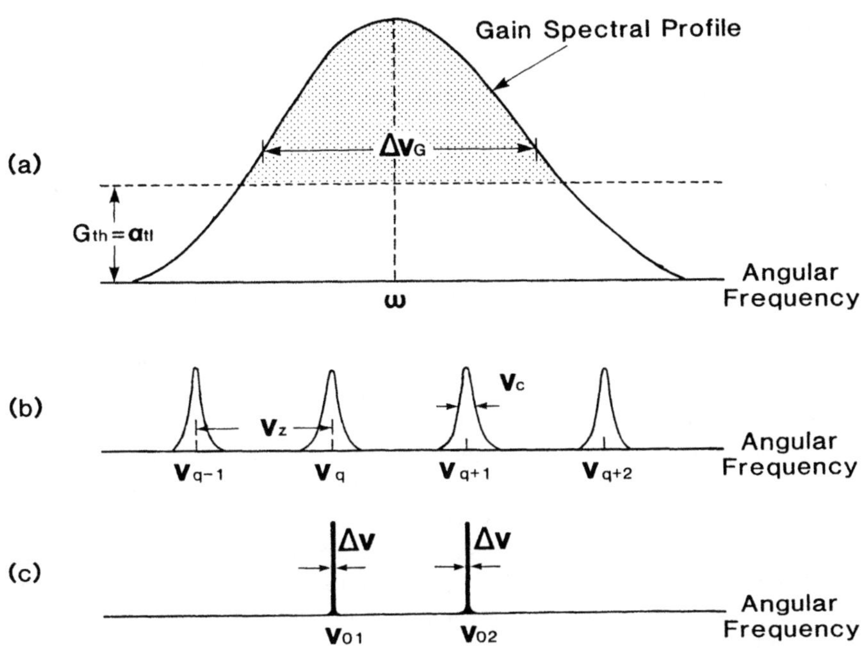

Fig. 2.4. Schematic explanation of the field spectral profile of laser oscillation [D]. (a) Gain spectral profile. (b) Resonance spectral profile of the longitudinal modes of the cavity. (c) Field spectral profile of laser oscillation. (FWHM $\Delta\nu$ is 2π times the Δf in Eq.(4-19)).

frequency is determined by the longitudinal mode frequency of the cavity falling within the gain spectral width, as shown in Figs. 2-4(b) and (c). However, as shown in Fig. 2-4(c), the field spectral linewidth of the laser oscillation $\Delta\nu$, representing the magnitude of frequency fluctuations, is much narrower than the widths of the gain spectrum $\Delta\nu_G$ and longitudinal mode spectrum of the cavity $\Delta\nu_c$. It depends on the magnitude of the random fluctuations of spontaneous emission.

2.3 Formulation of Laser Oscillation

Most parts of laser oscillation characteristics can be quantitatively described by the semi-classical theory, in which the behaviors of atoms and light are treated by quantum mechanics and classical wave theory, respectively. The method of semi-classical analysis of laser oscillation is described in this section [B].

2.3.1 Density matrix

In most of the discussion in 2.1, except for the derivation of the Einstein A coefficient, the interaction between one electron in an atom and light was treated. However, since a great number of atoms is confined in the actual laser cavity, the ensemble average of electron behaviors in these atoms should be used to formulate laser oscillation characteristics. Furthermore, the excitation of electrons by an external power source, relaxation, de-excitation by atomic collisions, and so on, should be included in this formulation. For this purpose, a method of statistical mechanics is employed by using a density matrix operator. This operator is defined by

$$\rho \equiv \sum_{\psi} P_{\psi} |\psi\rangle\langle\psi| \qquad (2-42)$$

where $|\psi\rangle$ is the atomic wavefunction defined by Eq.(2-22). The quantity P_ψ is the probability of finding that each atom in the ensemble stays in the state described by $|\psi\rangle$. This probability should be used in the above definition because the wavefunction of each atom is not exactly known as $|\psi\rangle$, i.e. in most case, only the value of this kind of probability can be known. Such a state of atomic ensemble is called a mixed state. On the other hand, i.e. in a special case, if the wavefunction of each atom is known to be $|\psi\rangle$ (the values of $c_n(t)$ in Eq.(2-22) are known exactly), the state of atomic ensemble is called a pure state. However, it is convenient to note that the density matrix operators for mixed and pure states are commonly defined by Eq.(2-42).

The ensemble average $\langle A\rangle$ of the quantum mechanical expectation value $\langle A\rangle$ ($\equiv \langle\psi|A|\psi\rangle$) of the arbitrary operator A can be given by using the density matrix operator, which is expressed as

$$\langle A\rangle = \mathrm{Tr}(\rho A) \qquad (2-43)$$

The operators ρ and A in the parenthesis of this equation are expressed by using matrices. The operator ρ expressed in the matrix notation is called the density matrix. The notation Tr in this equation represents the trace of the matrix in the parenthesis. Since P_ψ and c_n are normalized, it is found that

$$\mathrm{Tr}(\rho) = 1. \qquad (2-44)$$

The temporal variation of the density matrix is derived from Eqs.(2-20), (2-22) and (2-42), and is given by

$$\mathrm{d}\rho / \mathrm{d}t = \frac{i}{\hbar}\left[\rho, \mathcal{H}_0 + \mathcal{V}\right]. \qquad (2-45)$$

In the above discussions, only the behavior of the two-level atoms was treated under the perturbation of Eq.(2-19). However, practical atoms have a number of energy levels in addition to the two levels relevant to the present discussions. In these atomic ensembles, de-excitation from one level to the other due to atomic collisions, excitation, etc., takes place incoherently with a phase of incident light. Therefore, it is quite difficult by conventional formulation to trace the temporal variation of each atomic wave-function due to these incoherent perturbations, i.e. interactions with the reservoir. However, the effects due to relaxation and attenuation by de-excitations and to excitation in population inversions, can be phenomenologically introduced into Eq.(2-45) if the atomic ensemble is described by using a density matrix.

By noting that the diagonal elements ρ_{aa}, ρ_{bb} and the non-diagonal element ρ_{ab} which correspond to the population density of the upper and lower levels and the magnitude of the induced dipole moment in Eq.(2-31), respectively, the incoherent effects described above can be introduced into Eq.(2-45) and are expressed as

$$\frac{\mathrm{d}\rho_{aa}}{\mathrm{d}t} = -\gamma_a\left(\rho_{aa} - \rho_{aa}^0\right) + \frac{i}{\hbar}\left(\rho_{ab}\,\mathcal{V}_{ba} - \mathcal{V}_{ab}\,\rho_{ba}\right), \qquad (2-46)$$

$$\frac{\mathrm{d}\rho_{bb}}{\mathrm{d}t} = -\gamma_b\left(\rho_{bb} - \rho_{bb}^0\right) + \frac{i}{\hbar}\left(\rho_{ba}\,\mathcal{V}_{ab} - \mathcal{V}_{ab}\,\rho_{ba}\right), \qquad (2-47)$$

and

$$\frac{\mathrm{d}\rho_{ab}}{\mathrm{d}t} = -\left(i\omega + \gamma\right)\rho_{ab} + \frac{i}{\hbar}\mathcal{V}_{ab}\left(\rho_{aa} - \rho_{bb}\right), \qquad (2-48)$$

where γ_a and γ_b represent the relaxation time constants in the decrease of population densities of the upper and lower levels (due to de-excitations) to other energy levels, respectively. These time constants are called longitudinal relaxation constants which originate from the analogy with the formulation of nuclear magnetic resonance. The quantity γ represents the time constant of the decrease of the ensemble averaged induced dipole moment, which depends on the rate of decrease of the induced dipole moment in each atom and the rate of dephasing between these induced dipole moments. This is called the transverse relaxation constant. The quantities $\rho_{aa}{}^0$ and $\rho_{bb}{}^0$ represent stationary values of population densities of the upper and lower levels in the absence of the perturbation $\mathcal{V}$, respectively. They depend on the magnitude of energy supplied from the external power source, and the condition of $\rho_{aa}{}^0 > \rho_{bb}{}^0$ is usually satisfied since the population inversion is required for laser oscillation.

2.3.2 Van der pol's equation

To derive the equations describing temporal variations in the electric field vector of the laser light, a semi-classical method is employed [B]. As is schematically shown in Fig. 2-5, the procedure of derivation is:

(1) Derive the magnitude of the dipole moment $\langle ex \rangle$ induced in each atom by an assumed electric field of light, $E(z, t)$, in the cavity according to the law of quantum mechanics.

(2) Derive the magnitude of the induced polarization $P(z, t)$ of the atomic ensemble by summarizing these induced dipole moments in atoms with the help of a density matrix formulation.

(3) Derive the electric field of the secondary light $E'(z, t)$ by substituting $P(z, t)$ into Maxwell's equation. In this case, the induced polarization $P(z, t)$ acts as the source in generating the secondary field $E'(z, t)$, as the result of a reaction from the atomic ensemble to the incident field $E(z, t)$.

(4) The electric field of light is finally determined by applying the field self-consistency condition, i.e. the field $E(z, t)$ which induces polarization in the atomic ensemble, should be equal to the resulting field $E'(z, t)$. The electric field derived by such a method is called a self-consistent field.

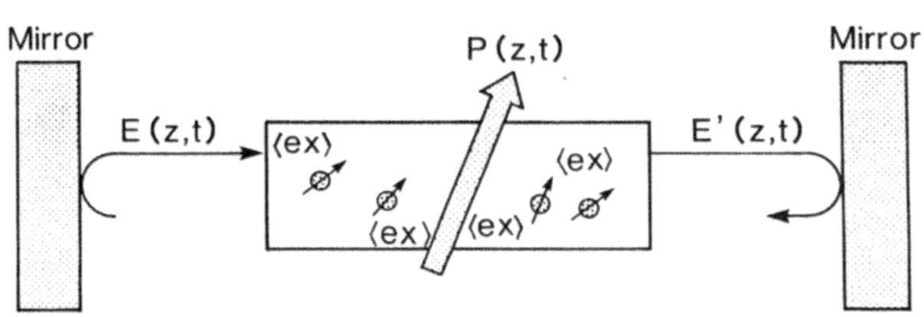

Fig. 2.5. Schematic explanation of semi-classical laser theory. $E(z, t)$ is the electric field incident into the active medium, $\langle ex \rangle$ is the induced dipole momentum, $P(z, t)$ is the induced polarization, $E'(z, t)$ is the electric field of the emitted light generated by $P(z, t)$.

[1] Derivation of induced polarization

The above-mentioned steps (1) and (2) are carried out by using the density matrix given in 2.3.1. If the electric field $E(z, t)$ in the cavity (Fig. 2-2(b)) is linearly polarized along the x-axis, one can ignore the vectorial character of the field. The electric field is then expressed as the sum of longitudinal modes in the cavity:

$$E(z\ t)=\frac{1}{2}\sum_q E_q(t)\exp\left(-i\nu_{ql}t\right)U_q(z)+c\ c \qquad (2-49)$$

This is the same as in Eq.(1-8), but the subscript x has been omitted for simplicity. The amplitude of the q-th mode $E_q(t)$ varies little in an optical frequency period and ν_{ql} is the oscillation angular frequency of the q-th mode. This angular frequency ν_{ql} is not necessarily equal to the passive cavity resonance frequency ν_q due to the dispersion of the active medium, or the atomic ensemble. The value of ν_{ql} falls within a range of about $\omega - \Delta\nu_G/2 \sim \omega + \Delta\nu_G/2$ in the gain spectral profile of Fig. 2-4(a) and within a range of about $\nu_C - \Delta\nu_c/2 \sim \nu_c + \Delta\nu_c/2$ in the passive cavity resonance spectral profile. The normal mode function $U_q(z)$ represents the spatial profile of the q-th mode along the z-axis. In the case of a standing wave in the Fabry-Perot cavity, it is expressed as

$$U_q(z)=\sin k_q z \qquad (2-50)$$

where k_q is the wave number and is expressed by the wavelength λ_q of Eq.(2-7) as

$$k_q = 2\pi / \lambda_q \qquad (2-51)$$

Similar to the form of the field $E(z, t)$ given by Eq.(2-49), the polarization of the atomic ensemble induced by $E(z, t)$ in Eq.(2-49) can be given by the expression

$$P(z\ t)=\frac{1}{2}\sum_q P_q(t)\exp\left(-i\nu_{ql}t\right)U_q(z)+c\ c \qquad (2-52)$$

If the value of the amplitude $E_q(t)$ in Eq.(2-49) is assumed as real, the amplitude $P_q(t)$ is, in general, complex because the phases of $E(z, t)$ and $P(z, t)$ are different. Since the value of $P(z, t)$ is directly derived by a density matrix, as will be shown later, $P_q(t)$ is derived as a projection of $P(z, t)$ onto the q-th Fourier component which is expressed as

$$P_q(t)=2e^{i\nu_{ql}t}\frac{1}{U_0}\int_0^L U_q^*(z)P(z\ t)\mathrm{d}z \qquad (2-53)$$

The normalization factor U_0 in this equation is given by

$$U_0 \equiv \int_0^L \left|U_q(z)\right|^2 \mathrm{d}z = L/2. \qquad (2-54)$$

Since the operator corresponding to the dipole moment induced in each atom is ex, the value of $P(z, t)$ is derived from Eq.(2-43) as

$$P(z,t) = \mathrm{Tr}(\rho ex) = p^* \rho_{ab}(z,t) + c.c., \qquad (2-55)$$

where p is the magnitude of the electric dipole matrix element defined by Eq.(2-26). Therefore, from Eqs.(2-53) and (2-55), one has

$$P_q(t) = 2p^* e^{i\nu_{ql}t} \frac{1}{U_0} \int_0^L U_q^*(z)\rho_{ab}(z,t)\mathrm{d}z. \qquad (2-56)$$

In the following discussions, a single mode oscillation is assumed for simplicity and a steady state solution is derived. In such a case, the summations on the right-hand sides of Eqs.(2-49) and (2-52) can be simplified and the magnitude of the perturbation energy matrix element $\mathcal{V}_{ab}$ of Eq.(2-48) is derived from Eqs.(2-19) and (2-49) and is expressed as

$$\mathcal{V}_{ab} = -\frac{p}{2}\left\{E_q(t)e^{-i\nu_{ql}t} + c.c.\right\}\cdot U_q(z)$$

$$\cong -\frac{p}{2} E_q(t)e^{-i\nu_{ql}t} U_q(z), \qquad (2-57)$$

in which a rotating-wave approximation is used for the final reduction of the equation. By assuming that the magnitudes of $E_q(t)$ in this equation and the population inversion density $\rho_{aa} - \rho_{bb}$ vary little within the time range of $1/\gamma$(= several ns) i.e. that energy is supplied nearly stationarily from the external power source, one has from Eq.(2-48)

$$\rho_{ab} = -\frac{i}{2\hbar} pE_q e^{-i\nu_{ql}t} U_q(z) \frac{\rho_{aa} - \rho_{bb}}{i(\omega - \nu_{ql}) + \gamma}. \qquad (2-58)$$

For the steady state d/dt = 0, the magnitude of the population inversion density is derived by substituting Eq.(2-58) into Eqs.(2-46) and (2-47), which is expressed as

$$\rho_{aa} - \rho_{bb} = \frac{N(z)}{1 + R(z)/R_s}, \qquad (2-59)$$

where

$$R(z) = \frac{1}{2\gamma}\left(\frac{|p|E_q}{\hbar}\right)^2 |U_q(z)|^2 \mathcal{L}(\omega - \nu_{ql}), \qquad (2-60)$$

in which the quantity $R(z)$ is the rate constant in the variation of the population density of each energy level due to stimulated emission and is proportional to the Einstein B coefficient of Eq.(2-30). $\mathcal{L}$ represents the Lorentzian defined by

$$\mathcal{L}(x) \equiv \frac{\gamma^2}{x^2 + \gamma^2}. \qquad (2-61)$$

This function already appeared in Eq.(2-14). Furthermore, R_s is called the saturation parameter, defined by

$$R_s \equiv \frac{\gamma_a \gamma_b}{\gamma_a + \gamma_b}. \qquad (2-62)$$

The unsaturated population inversion density $N(z)$ is defined by

$$N(z) \equiv \rho_{aa}^0 - \rho_{bb}^0, \qquad (2-63)$$

which represents the magnitude of the population inversion density in the absence of incident light and is determined by energy supplied from the external power source and the relaxation rates. For the rate constant R with $|U_q(z)|^2 = \sin^2 kz$ dependence, the population inversion density of Eq.(2-59) also shows this dependence. More specifically, the sinusoidally varying electric field intensity burns holes in the distribution of the population inversion density along the z-axis, spaced half a wavelength apart, as shown in Fig. 2-6. This phenomenon is called spatial hole burning and is the result of stimulated emission at the anti-nodes of standing wave. By using Eq.(2-56) after substituting Eq.(2-59) into Eq.(2-58), one has

$$P_q(t) = -\frac{|p|^2 E_q(t)}{\hbar} \frac{i}{i(\omega - \nu_{ql}) + \gamma} \frac{1}{U_0} \int_0^L \frac{|U_q(z)|^2 N(z)}{1 + R(z)/R_s} dz. \qquad (2-64)$$

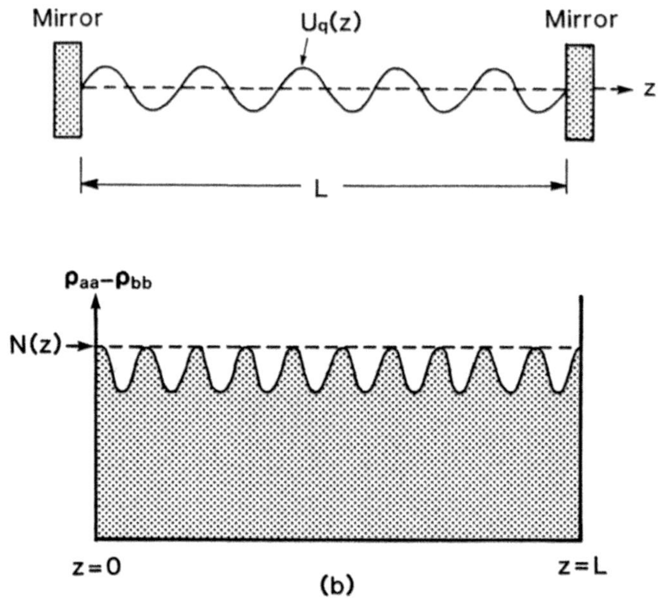

Fig. 2.6. (a) Spatial profile of the standing wave in cavity with an amplitude of $U_q(z)$ (see Eq.(2-50)). (b) Spatial hole-burning of the population inversion.

This is the result of the present discussion. Since secondary light is generated from this induced polarization, the spectral profile of this polarization gives a gain spectral profile for laser oscillation. It is easily seen from this equation that the gain spectral profile has a Lorentzian shape $\mathcal{L}(\omega-\nu_{ql})$ with a half width at half maximum (HWHM) of γ. This profile corresponds to the curve in Fig. 2-4(a). Within the limit of low optical power, the approximation

$$1/\left\{1+R(z)/R_s\right\} \cong 1-\frac{R(z)}{R_s}. \tag{2-65}$$

holds. The integration in Eq.(2-64) can easily be carried out under this approximation, which gives

$$P_q(t) = -\frac{|p|^2 E_q(t)}{\hbar}\frac{i}{i\left(\omega-\nu_{ql}\right)+\gamma}\overline{N}\left[1-\frac{3}{8\gamma R_s}\left(\frac{|p|E_q(t)}{\hbar}\right)^2 \mathcal{L}\left(\omega-\nu_{ql}\right)\right]$$

$$\equiv P_q^{(1)}(t)+P_q^{(3)}(t), \tag{2-66}$$

The population inversion density averaged over the cavity length is given by

$$\overline{N} \equiv \frac{1}{U_0}\int_0^L N(z)\left|U_q(z)\right|^2 \mathrm{d}z \cong \frac{1}{L}\int_0^L N(z)\mathrm{d}z, \qquad (2-67)$$

where it was assumed that the magnitude of $N(z)$ varies little in the period of $U_q(z)$. Equation (2-66) represents an approximation of the third order with respect to the amplitude E_q, which is sufficiently accurate to describe many laser oscillation characteristics. The theoretical analysis under this approximation is called a third-order theory, where

$$P_q^{(j)}(t) \propto \left\{E_q(t)\right\}^j \qquad (j = 1, 3). \qquad (2-68)$$

If the amplitude coefficient $P_q(t)$ of Eq.(2-66) is expressed by using the electrical susceptibility χ_q, one has

$$P_q(t) = \varepsilon_0 \chi_q E_q(t) = \varepsilon_0\left(\chi'_q + i\chi''_q\right)E_q(t). \qquad (2-69)$$

By comparing this relation with Eq.(2-66), it is easily seen that the value of the imaginary part χ_q" is negative, i.e. the phase of the polarization delays that of the incident electric field as shown in the discussion of Eq.(2-32). Furthermore, by Eq.(2-66), χ_q, can be expressed as

$$\chi_q \cong \chi_q^{(1)} - \chi_q^{(3)} E_q^2(t). \qquad (2-70)$$

The quantity $\chi^{(3)}$ in this equation is called the third-order nonlinear susceptibility.

[2] Determination of the electric field

Steps (3) and (4) are carried out here, i.e. the amplitude coefficient of the electric field is derived by substituting Eq.(2-69) into Maxwell's equations. Since the secondary electric field is equal to the initially assumed incident field because of the demand on self-consistency, Eq.(2-49) is valid for both electric fields.

Maxwell's equation in vacuum (see Eqs.(1-2) ~ (1-7)) should be modified here in order to derive the electric field in the gain medium for laser oscillation. For this modification, Eqs.(1-2) and (1-7) are replaced by

$$\nabla \times \boldsymbol{H} = \partial \boldsymbol{D} / \partial t + \sigma \boldsymbol{E}, \qquad (2-71)$$

$$\boldsymbol{D} = \varepsilon_0 \boldsymbol{E} + \boldsymbol{P}. \qquad (2-72)$$

The conductivity σ in Eq.(2-71) was introduced to avoid a complicated boundary condition problem in the cavity and gives the damping due to absorption, scattering, diffraction and transmission of the cavity reflector. The polarization $\boldsymbol{P}$ in Eq.(2-72) has been derived from Eq.(2-69). The equation only for $\boldsymbol{E}$ and $\boldsymbol{P}$ is derived from Eqs.(1-3) ~ (1-6) and (2-71) and is expressed as

$$\nabla \times \nabla \times \boldsymbol{E} = -\mu_0 \sigma \frac{\partial \boldsymbol{E}}{\partial t} - \mu_0 \varepsilon_0 \frac{\partial^2 \boldsymbol{E}}{\partial t^2} - \mu_0 \frac{\partial^2 \boldsymbol{P}}{\partial t^2} \tag{2-73}$$

The left side of this equation can be transformed by using the formulae of vector analysis under the approximation of $\partial/\partial x$ and $\partial/\partial y \cong 0$ because the variations in the field intensity transverse to the z-axis vary slowly compared to those along the z-axis in the cavity. Furthermore, the z-components of $\boldsymbol{E}$ and $\boldsymbol{P}$ are neglected because light can be assumed as a transverse wave. Under these approximations, Eq.(2-73) can be reduced to

$$-\frac{\partial^2 \boldsymbol{E}}{\partial z^2} + \mu_0 \sigma \frac{\partial \boldsymbol{E}}{\partial t} + \mu_0 \varepsilon_0 \frac{\partial^2 \boldsymbol{E}}{\partial t^2} = -\mu_0 \frac{\partial^2 \boldsymbol{P}}{\partial t^2} \tag{2-74}$$

The projection onto the q-th mode (see Eq.(2-53)) after substituting Eqs.(2-49) and (2-52) into this equation gives

$$\nu_q^2 E_q - i\left(\frac{\sigma}{\varepsilon_0}\right)\nu_{ql} E_q - 2i\nu_{ql}\frac{dE_q}{dt} - \nu_{ql}^2 E_q = \frac{\nu_{ql}^2}{\varepsilon_0} P_q \tag{2-75}$$

Here, the terms containing $\partial^2 E_q/\partial t^2$, $\partial^2 P_q/\partial t^2$, $\sigma \partial E_q/\partial t$ and $\partial P_q/\partial t$ were neglected because $E_q(t)$ and $P_q(t)$ vary little in an optical frequency period, the losses are small and the medium is diluted. Since the decrease in optical energy is given by Eq.(2-11), the conductivity σ representing the cavity loss can be derived from Eqs.(2-11), (2-13) and (2-16) and is expressed as

$$\sigma = \alpha'_{\mathrm{tl}}\, \varepsilon_0 = \frac{\varepsilon_0 \nu_{ql}}{Q_q} \tag{2-76}$$

By noting that the value of P_q is small due to a diluted gain medium and $|(\omega - \nu_q)/\omega| \leq \gamma/\omega << 1$ because the linewidth of the spectral profile of P_q is γ, one can obtain reduced expressions for the real and imaginary parts of Eq.(2-75); they are

$$\frac{\mathrm{d}E_q}{\mathrm{d}t} + \frac{\omega}{2Q_q} E_q = -\frac{\omega}{2\varepsilon_0} \mathrm{Im}\left[P_q\right] \tag{2-77}$$

and

$$\nu_{ql} = \nu_q - \frac{\omega}{2\varepsilon_0 E_q}\mathrm{Re}\left[P_q\right], \tag{2-78}$$

where Im[ ] and Re[ ] represent imaginary and real parts of the quantity in brackets, respectively. These two equations give expressions for the amplitude coefficient $E_q(t)$ and the angular frequency ν_{ql} of a laser oscillation, respectively. In the absence of gain medium (i.e. $P_q = 0$), Eq.(2-77) gives $E_q \propto \exp(-\omega t/2Q_q)$ which corresponds to the expression given by Eq.(2-13). Furthermore, in that case Eq.(2-78) would give $\nu_{ql} = \nu_q$, i.e. the field angular frequency which coincides with the resonance angular frequency of the empty cavity. However, when a gain medium exists, the substitution of Eq.(2-69) into Eq.(2-77) gives

$$\mathrm{d}E_q / \mathrm{d}t = -\frac{\omega}{2Q_q}E_q - \frac{\omega}{2}\chi''_q E_q. \tag{2-79}$$

Since χ_q" takes on a negative value, the second term on the right-hand side of this equation represents the value of E_q increasing with time, i.e. $-\chi_q$" represents the gain in laser oscillation. On the other hand, the substitution of Eq.(2-69) into Eq.(2-78) gives

$$\nu_{ql} = \nu_q - (\omega / 2)\chi'_q, \tag{2-80}$$

which means that the angular frequency of a laser oscillation deviates from the resonance angular frequency of the empty cavity. Since the refractive index of the medium is expressed as

$$n\left(\nu_{ql}\right) \equiv \nu_q / \nu_{ql} \cong 1 + \frac{1}{2}\chi'_q, \tag{2-81}$$

it is found that χ_q' corresponds to the refractive index of the gain medium for light with an angular frequency of ν_{ql}.

Furthermore, substitution of Eq.(2-69) into Eq.(2-77) gives

$$\frac{\mathrm{d}E_q}{\mathrm{d}t} + \left(-\frac{\omega}{2Q_q} + \frac{\omega}{2}\chi_q^{(1)''} - \frac{\omega}{2}\chi_q^{(3)''}E_q^2\right)E_q = 0, \tag{2-82}$$

which is a nonlinear differential equation because it has the term proportional to E_q^2

in parentheses. This is called a Van der Pol equation, which has been well used to describe the variation in field amplitude not only in lasers but also in general self-sustained oscillators.

[3] Steady-state values of the amplitude coefficient and angular frequency

The substitution of Eq.(2-69) into Eq.(2-77) gives

$$\frac{dE_q}{dt} = E_q\left(-\Gamma_q + \alpha_q - \beta_q E_q^2\right), \tag{2-83}$$

which is also a Van der Pol equation; however, further detailed informations can be extracted from this expression than from Eq.(2-82). Here, Γ_q, α_q and β_q are called the cavity loss, linear gain (or small signal gain) and the gain saturation coefficient, respectively, and are expressed as

$$\Gamma_q = \omega / 2Q_q, \tag{2-84}$$

$$\alpha_q = \frac{\omega |p|^2 \overline{N}}{2\varepsilon_0 \hbar \gamma} \mathcal{L}\left(\omega - \nu_{ql}\right), \tag{2-85}$$

and

$$\beta_q = \frac{3\omega |p|^4 \overline{N}}{16\varepsilon_0 \hbar^3 \gamma^2 R_s} \left\{\mathcal{L}\left(\omega - \nu_{ql}\right)\right\}^2. \tag{2-86}$$

The linear gain α_q represents a gain proportional to the average density of population inversion $\overline{N}$ determined by the magnitude of the energy supplied from a power source. The gain saturation coefficient β_q represents the decrease of gain by the decrease of population inversion as a result of stimulated emission. By defining a quantity proportional to the laser power

$$I_q \equiv E_q^2, \tag{2-87}$$

one has from Eq.(2-83)

$$\frac{dI_q}{dt} = 2I_q\left(-\Gamma_q + \alpha_q - \beta_q I_q\right). \tag{2-88}$$

Under the initial condition of $I_q = I_0$ caused by spontaneous emission at $t = 0$, the solution of this equation can be approximated as

$$I_q \cong I_0 \exp\left\{2\left(\alpha_q - \Gamma_q\right)t\right\}, \tag{2-89}$$

by neglecting the effect of gain saturation since laser power is low at around $t = 0$. This solution implies that laser power begins an exponential increase just after the external power supply has been turned on if $\alpha_q > \Gamma_q$. The effect of gain saturation can not be neglected any more if the laser power is increased with time and the steady state solution realized at the limit of $t \to \infty$ is derived by setting $d/dt = 0$. Then, one has

$$I_q = \left(\alpha_q - \Gamma_q\right) / \beta_q. \tag{2-90}$$

This shows that the steady state laser power is determined by the ratio of net gain $\alpha_q - \Gamma_q$ to the gain saturation coefficient β_q.

The threshold of laser oscillation is defined by setting the net gain $\alpha_q - \Gamma_q$ to zero at the center frequency ($\nu_{ql} = \omega$) of the gain spectral profile. By this definition, the population inversion density at the threshold is derived from Eqs.(2-84) and (2-85), and is expressed as

$$\overline{N}_{\text{th}} = \varepsilon_0 \hbar \gamma / |p|^2 Q_q. \tag{2-91}$$

Equation (2-90) can be transformed by this expression to

$$I_q = \frac{8\gamma R_s}{3} \cdot \left(\frac{\hbar}{|p|}\right)^2 \cdot \frac{\mathcal{L}\left(\omega - \nu_{ql}\right) - \overline{N}_{\text{th}} / \overline{N}}{\left\{\mathcal{L}\left(\omega - \nu_{ql}\right)\right\}^2}, \tag{2-92}$$

which means that, as shown in Fig. 2-7, laser power increases with an increase in $\overline{N}$, or an increase in energy supplied from an external power source.

The angular frequency is expressed, by using Eqs.(2-66) and (2-78), as

$$\nu_{ql} = \nu_q + \sigma_q - \rho_q I_q, \tag{2-93}$$

where σ_q and ρ_q are called the linear pulling and nonlinear pushing coefficients, respectively, and are shown as

$$\sigma_q = \frac{\omega |p|^2 \overline{N}}{2\varepsilon_0 \hbar} \frac{\omega - \nu_{ql}}{\gamma^2} \mathcal{L}\left(\omega - \nu_{ql}\right), \tag{2-94}$$

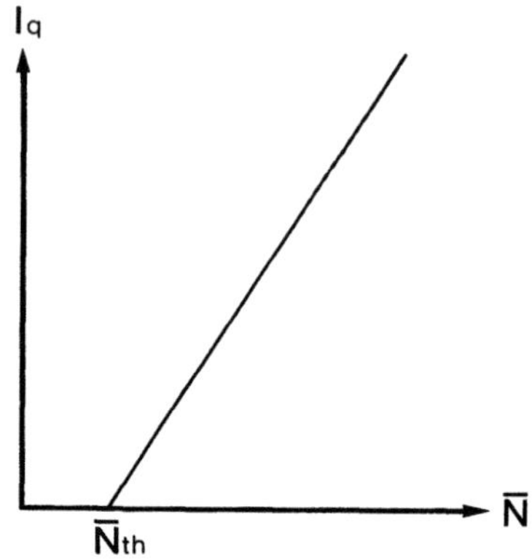

Fig. 2.7. Relation between steady state laser power I_q and average population inversion $\overline{N}$.

and

$$\rho_q = \frac{3\omega|p|^4 \overline{N}}{16\varepsilon_0 \hbar^3 \gamma^3 R_s}\left(\omega - \nu_{ql}\right)\left\{\mathcal{L}\left(\omega - \nu_{ql}\right)\right\}^2. \qquad (2-95)$$

Since $\mathcal{L}(\omega - \nu_{ql})$ varies slower than $(\omega - \nu_{ql})$ when ν_{ql} is around ω, these two equations can be approximated by $\sigma_q = \sigma_0 (\omega - \nu_{ql})$ and $\rho_q = \rho_0 (\omega - \nu_{ql})$, where σ_0 and ρ_0 are proportional constants determined by Eqs.(2-94) and (2-95). By substituting these approximated expressions into Eq.(2-93), one has

$$\nu_{ql} = \frac{\nu_q + \left(\sigma_0 - \rho_0 I_q\right)\omega}{1 + \left(\sigma_0 - \rho_0 I_q\right)}. \qquad (2-96)$$

By neglecting the nonlinear term $\rho_0 I_q$ which is smaller than the linear term σ_0, it can be easily found from this expression that the laser angular frequency is pulled from the resonance angular frequency ν_q of the empty cavity toward the center angular frequency ω of the gain spectral profile (compare Figs. 2-4(b) and (c)). This resulted from a change in refractive index in the cavity owing to the gain medium with a population inversion. The quantity $-\rho_0 I_q$ is the correction factor required in high laser power, which shows that the laser oscillation angular frequency ν_{ql} is pushed out from ω to ν_q. This is because the refractive index of the gain-cavity gets slightly closer to that of the empty-cavity again due to a decrease in population inversion by the stimulated emission in laser oscillation. The resultant frequency deviation due to linear pulling and nonlinear pushing can be estimated by substituting Eqs. (2-92), (2-94) and (2-95) into Eq.(2-93), which gives

$$\nu_{ql} = \frac{\nu_q + S\omega}{1+S} \qquad (2-97)$$

This equation can be interpreted as the mass-center equation and the oscillation angular frequency ν_{ql} can be regarded as the average value between ν_q and ω with weights 1 and S. The quantity S is called the stabilization factor and is expressed as

$$S = \frac{\omega^2 |p|^2 \overline{N}_{\text{th}}}{2\varepsilon_0 \hbar \gamma^2} \qquad (2-98)$$

This expression can be transformed by using Eqs.(2-15) and (2-91); consequently, one has

$$S = \left(\Delta\nu_c / 2\right) / \gamma \qquad (2-99)$$

This corresponds to the ratio of widths of the resonance cavity spectral profile to the width of the gain spectral profile. Since $S \ll 1$, the laser frequency ν_{ql} is much closer to ν_q than to ω, which means that the fluctuations of the laser angular frequency strongly depend on that of the cavity resonance frequency.

The formulation of multi-mode laser oscillation can also be carried out and equations corresponding to Eqs.(2-88) and (2-93) are derived by a similar procedure to that for the single mode oscillation described above. For the simplest multi-mode oscillation, i.e. a two-mode oscillation, they are

$$\mathrm{d}I_q / \mathrm{d}t = 2I_q\left(-\Gamma_q + \alpha_q - \beta_q I_q - \theta_{pq} I_p\right) \qquad (2-100)$$

and

$$\nu_{ql} = \nu_q + \sigma_q - \rho_q I_q - \tau_{qp} I_p \qquad (q \neq p) \qquad (2-101)$$

where θ_{qp} is called the cross saturation coefficient representing gain decrease in the q-th mode due to a decrease in population inversion induced as a result of oscillation in the p-th mode. The quantity τ_{qp} is called a cross pushing coefficient representing mutual pushing between the angular frequencies ν_{1l} and ν_{2l}. Cross-pushing occurs because the refractive index of the gain-cavity for the q-th mode approaches that of the empty-cavity, because the population inversion is decreased by an oscillation in p-th mode. Equation (2-100) implies that the additional power fluctuations, which appear in each mode during the multi-mode oscillation, are represented by the term proportional to θ_{qp}. Furthermore, Eq.(2-101) shows that the angular frequency fluctuations of each mode increase due to additional power fluctuations.

[Textbooks]

[A] K. Shimoda, T. Yajima, Y. Ueda, T. Shimizu and T. Kasuya, *Quantum Electronics,* Shokabo Publishing, Tokyo, 1972 (in Japanese).
[B] M. Sargent, III, M. O. Scully and W. E. Lamb, Jr., *Laser Physics,* Addison-Wesley, Reading, 1974.
[C] R. Loudon, *The Quantum Theory of Light,* Clarendon Press, Oxford, 1973.
[D] M. Ohtsu, "Technology for Light Emitting and Detecting Devices for Fiber Sensor", Chapter 3 in *Optical Fiber Sensor*, ed. by T. Okoshi, Ohm-sha Publishing, Tokyo, 1986 (in Japanese).

[References]

1) W. W. Rigrod, "Isolation of axi-symmetrical optical-resonator modes", *Appl. Phys. Lett.*, Vol. 2, (1983) pp. 51–53.

Chapter 3

Structures and Characteristics of Lasers

If laser devices are classified according to host materials, oscillation wavelengths, power and so on, many classes can be listed. Reviews in this chapter are limited only to representative examples of continuous-wave (CW) laser devices.

3.1 Gas Lasers

As representative gas lasers, helium-neon (He-Ne), argon ion (Ar^+) and carbon dioxide (CO_2) lasers are reviewed.

3.1.1 Helium-neon (He-Ne) laser

The He-Ne laser is a well-developed device and has been widely used because it is more compact than the other gas lasers and exhibits a stable CW oscillation with a red color (with a wavelength of 0.63 μm). Its typical structure is shown in Fig. 3-1. Its cavity has a length of about 20 cm, in which a glass capillary tube of about 10 cm long and with an inner diameter of about 1 mm is installed and filled with He and Ne gases. The pressure ratio and total pressure are about 5:1 and 3 Torr, respectively. A glow discharge of approximately 5 mA is obtained by applying a high voltage of about 1 ~ 3 kV. Optically flat glass plates are fixed on both edges of this tube with an angle of θ_B (= $\tan^{-1} n_G$, where n_G is the refractive index of the

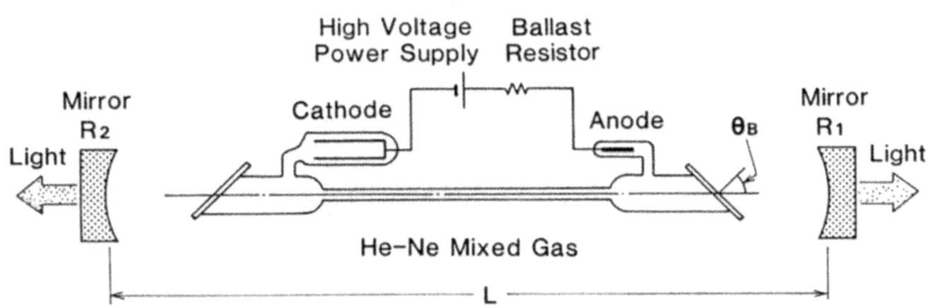

Fig. 3.1. Structure of a He-Ne laser [D].

glass). This angle is called the Brewster angle, at which glass plates have a zero reflection to the component linearly polarized along the plane of this figure, and thus, a linearly polarized laser oscillation is achieved.

Figure 3-2 shows the energy levels of He and Ne atoms at three wavelengths of laser oscillations (0.63 μm, 1.15 μm, and 3.39 μm) which can easily be oscillated by commonly using upper or lower energy levels. The oscillation at a 0.63 μm wavelength is suppressed by population depletion of the upper level due to the 3.39 μm wavelength oscillation. Cavity mirrors with a wavelength dependent reflectivity are used to realize a selective oscillation in order to overcome the problem of this sort of competition. Since these energy levels are located much higher than the ground state, they have a low population and, therefore, the magnitude of population inversion is small even though the energy is supplied from an external power source by discharge. Helium gas is used to increase population inversion through an excitation transfer from He gas to Ne gas, in which collisions between excited He atoms and ground state Ne atoms take place.

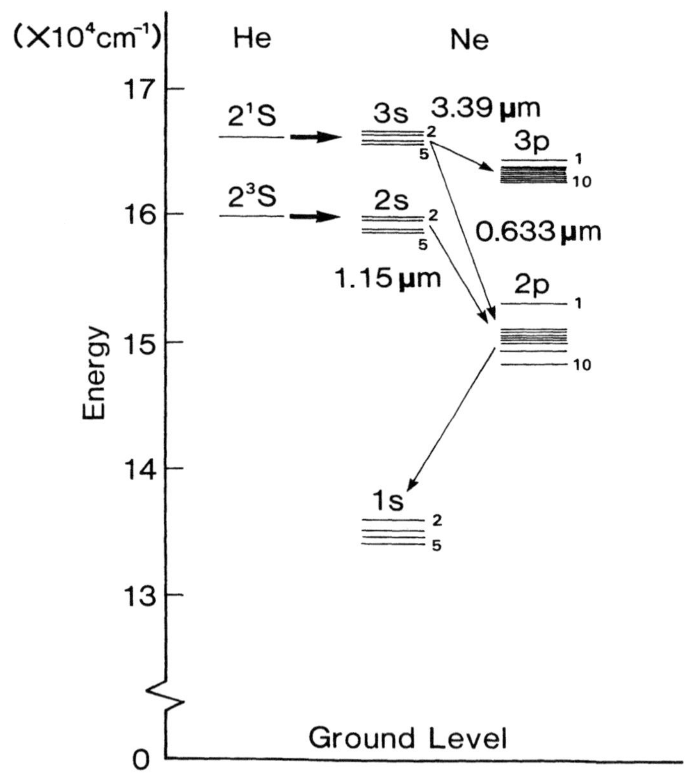

Fig. 3.2. Energy levels of electron in He and Ne atoms [A]. The typical lasing transitions are represented by arrows.

Discussions are now limited to the characteristics of a 0.63 μm wavelength line. Its linear gain is usually small since atomic density in the gas is low ($10^{15} \sim 10^{17}$ cm^{-3}), i.e. the linear gain α_q given by Eq.(2-85) is as small as $10^{-4} \sim 10^{-3}$ cm^{-1}, which requires a cavity with a very low loss in order to achieve laser oscillation. A stable cavity satisfying the condition of Eq.(2-1) is formed by using two mirrors with a surface flatness within about 50 nm. Mirror reflectivities are usually higher than 99%. By substituting, for example, $R_1 = R_2 = 99.8\%$ and $L = 20$ cm into Eq.(2-10), one obtains a cavity loss of $\alpha_{tl} = 2 \times 10^{-4}$ cm^{-1}, where the internal loss α_l is neglected because gas density is low and the lower energy level for laser oscillation is located far from the ground state. As long as the cavity loss is kept sufficiently low, CW laser oscillation will take place even though the linear gain is low. However, the present available steady laser power is only as low as 0.1 mW ~ 5 mW; this implies that the device efficiency, i.e. its ratio of optical output power to electrical input power, is as low as 0.01%.

Here, as an exception, one should note that the linear gain for the 3.39 μm wavelength line is much higher than that for the 0.63 μm, and is as high as 4 cm^{-1}. This is because the gain spectral profile has a narrow linewidth and excitation from the ground state to the lower level is prohibited. In addition, the excitation transfer from the metastable He level to the lower level is ineffective due to the large energy difference between them. However, even for this line, the steady power from Eq.(2-90) is low owing to the large gain saturation coefficient β_q from Eq.(2-86). For a high CW laser power, a small gain saturation coefficient β_q as well as a large linear gain α_q are necessary.

Equation (2-8) shows that the longitudinal mode separation $\Delta\nu_q/2\pi$ is 750 MHz in the present cavity, where the refractive index n_q of the gain medium (diluted He and Ne gases) was assumed as unity. The linewidth of the cavity resonance spectral profile and the cavity Q factor are 500 kHz and about 1×10^9, respectively, following Eqs.(2-15) and (2-16). The gain spectral linewidth $\Delta\nu_G/2\pi$ is about 1 GHz ~ 2 GHz, determined by the Doppler broadening caused by the Doppler effect due to thermal motions of Ne atoms at an electronic temperature of 500 °C in the discharge tube. Since the gain spectral width $\Delta\nu_G$ is about twice the longitudinal mode separation $\Delta\nu_q$, the adjacent two or three longitudinal modes can oscillate simultaneously. For transverse modes, only the lowest mode can be oscillated by installing a small aperture in the cavity or by adjusting the curvature radius of the two concave mirrors. The divergence angle of Eq.(2-5) is about 2 mrad for, e.g., the confocal cavity, which means that a very sharp directivity is obtained. The spot size on the surface of the mirror is estimated from Eq.(2-4) as 0.15 mm.

The laser power or frequency can be modulated by using a piezo-electric transducer (PZT) fixed on a cavity mirror or an intra-cavity electro-optical modulator (E/O modulator). The available maximum modulation frequencies are about 100 kHz and 100 MHz by the PZT and E/O modulator, respectively.

3.1.2 Argon ion (Ar^+) laser

Although the Ar^+ laser is more complicated in its structure, higher in price and shorter in lifetime than the He-Ne laser, it has been popularly used because of its high power CW oscillation in the visible wavelength region. An example of the device is shown in Fig. 3-3, where Ar atoms are ionized by electron collision generated by a large discharge current I. Schematic explanations of energy levels relevant to these laser oscillations are given in Figs. 3-4(a) and (b). A CW laser oscillation is achieved by a two-step excitation as shown in Fig. 3-4(a). In other words, atoms are ionized and excited to the upper level in laser oscillation by the first and second electron collision, respectively. Due to this two-step excitation, the linear gain in laser oscillation and the spontaneous emission power are proportional to I^2 and the steady-state laser power is proportional to I^4 [1]. There are several lasing lines ranging from green to violet, among which the most popular is that with a 514.5 nm wavelength.

The basic device structure of the Ar^+ laser is similar to that of the He-Ne laser; however, the discharge current density and the power per unit length of the discharge tube are as high as about 500 A/cm^2 and 50 W/cm, respectively. Due to the sputtering effect of the inner wall of the discharge tube by ionized atoms, the Ar gas pressure will gradually decrease. To solve this problem, several techniques have been employed such as: a discharge tube made of heat-proof materials (e.g., BeO) and with a water-cooled outer wall; a discharge tube connected to the gas reservoir in order to maintain the gas pressure; and an axial magnetic field applied to the discharge tube to avoid a decrease in gain induced by plasma diffusion and to protect the inner wall of the tube.

One oscillation line among those in Fig. 3-4(a) is selected by adjusting the angle of the intra-cavity prism, as shown in Fig. 3-3. For each line, a multi-longitudinal mode oscillation can usually be seen because the gain spectral linewidth is as large as 5 GHz. To obtain a single longitudinal mode oscillation, a Fabry-Perot etalon (the optical element same as the Fabry-Perot interferometer shown in Figs. 2-2(b) and 5.7) is also installed in the cavity.

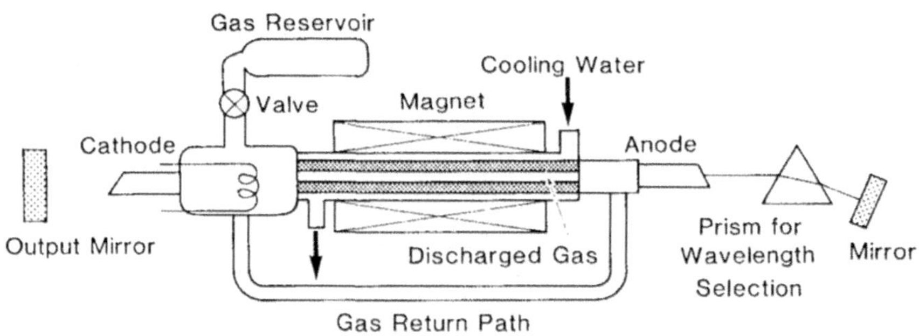

Fig. 3.3. Configuration of an argon ion laser [C].

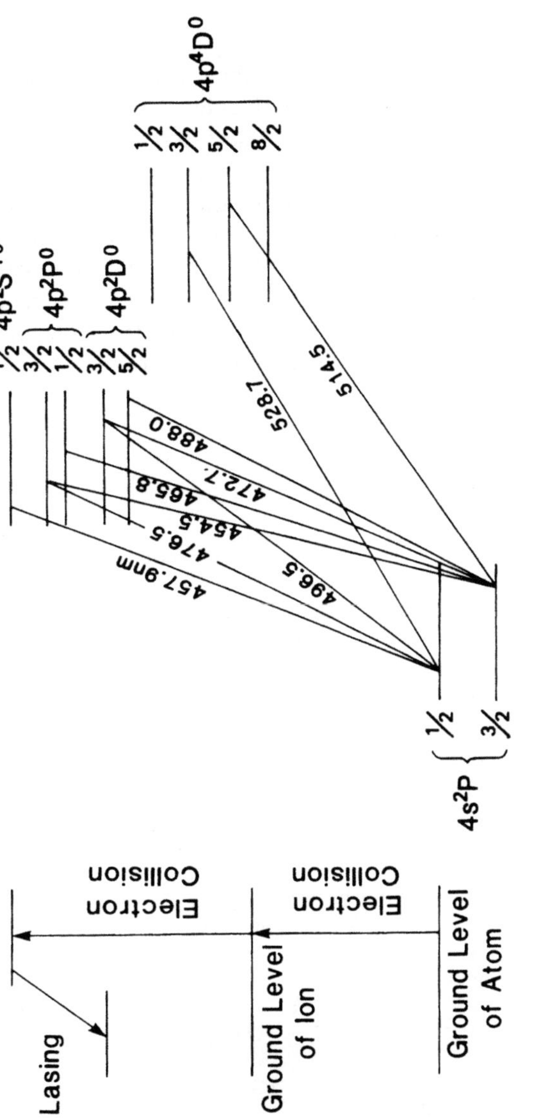

Fig. 3.4. Energy levels of electrons in an argon ion [A]. (a) Simplified description. (b) Detailed description to represent the typical lasing transitions.

The average value of the laser power is about 1 ~ 20 W for multi-line oscillation, i.e., without using a prism; however, a maximum power as high as over 100 W can also be obtained. A Krypton ion (Kr^+) laser with a wavelength range from about 340 nm to 800 nm operates on almost the same mechanism as that of an Ar^+ laser.

3.1.3 Carbon dioxide (CO_2) laser

A CO_2 laser oscillation is realized by using the transition between the energy levels of vibration-rotation motions of CO_2 molecules. Due to this lasing mechanism, the CO_2 laser is different from the other lasers described in 3.1.1 and 3.1.2, in which atomic electron transitions were used. A great number of CW oscillation lines can be obtained in the infrared wavelength region between 9 ~ 10 μm. Since the output power is as high as 1 W–10 kW, it has been popularly used for a variety of applications such as laser machining, welding, material processing, optical pumping for far-infrared lasers, analytical spectroscopy and laser radar. Typical structures of the laser device are shown in Figs. 3-5(a) and (b), which represent the conventional structure and the hollow waveguide-type structure for reducing the device dimension, respectively.

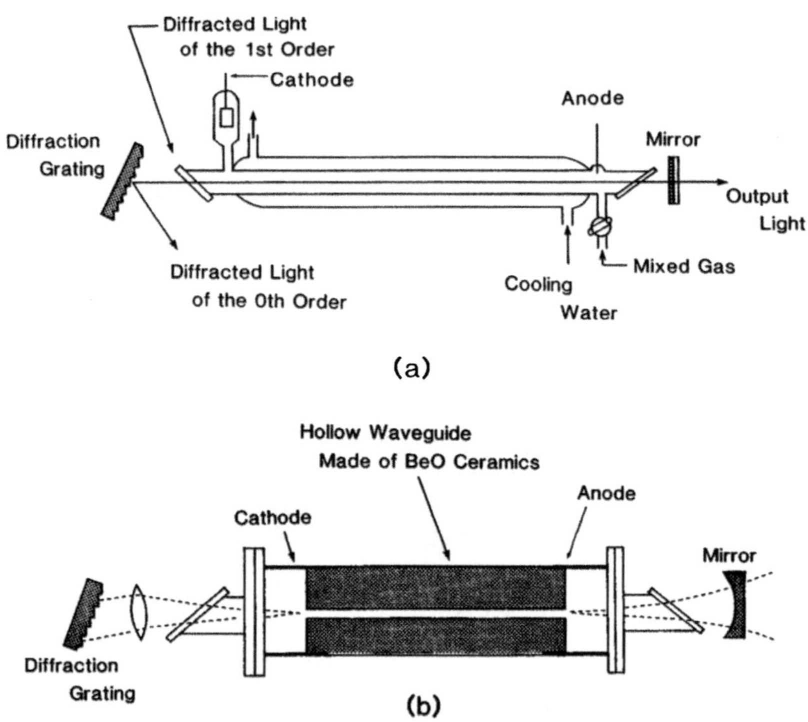

Fig. 3.5. Configuration of a carbon dioxide laser. (a) Conventional type [A]. (b) Hollow waveguide type.

Figures 3-6(a) and (b) show the energy levels of a CO_2 molecule relevant to laser oscillation. Figure 3-6(a) illustrates principal vibrational energy levels and wavelengths, i.e., 10.6 and 9.6 μm transitions. N_2 gas is mixed with CO_2 gas to excite CO_2 molecules efficiently through the excitation transfer from excited N_2 molecules to CO_2 molecules. Part of Fig. 3-6(a), as relating to the energy levels relevant to the 10.6 μm transition, is specified and shown in Fig. 3-6(b) which shows that each

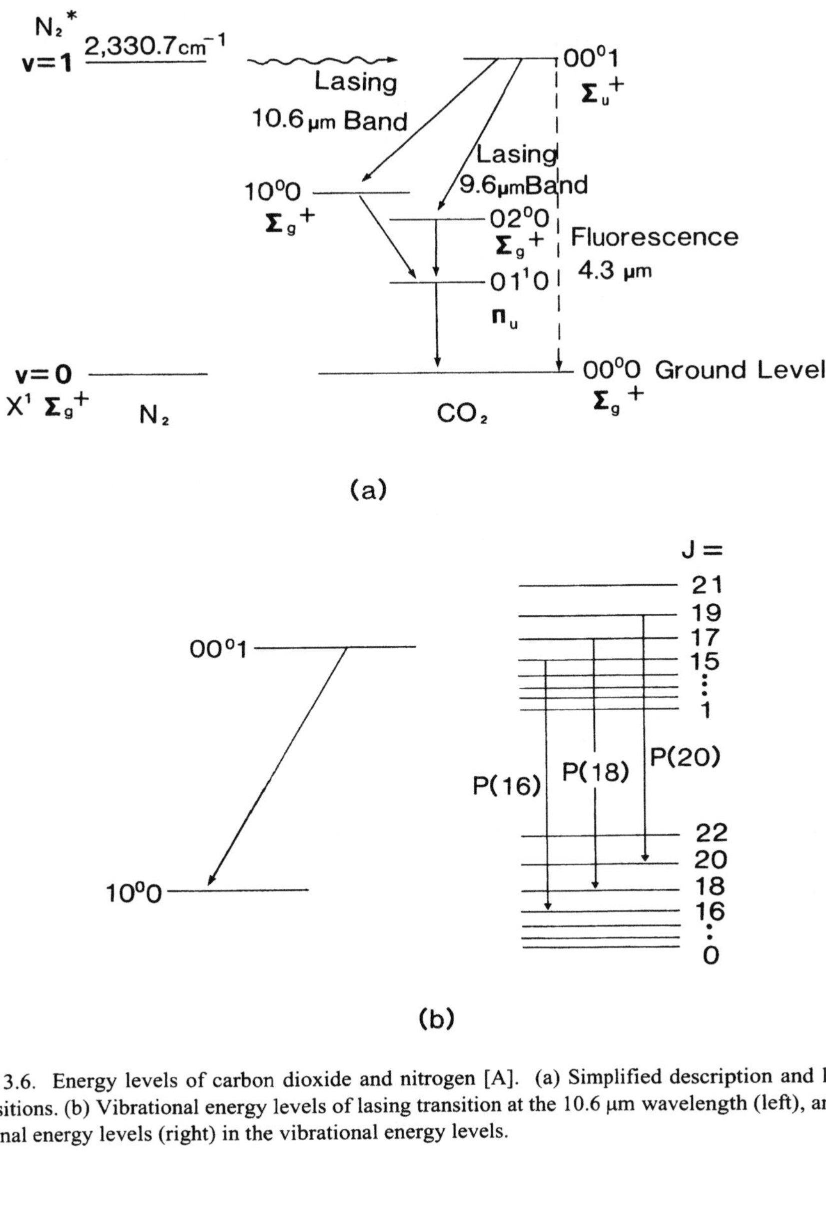

Fig. 3.6. Energy levels of carbon dioxide and nitrogen [A]. (a) Simplified description and lasing transitions. (b) Vibrational energy levels of lasing transition at the 10.6 μm wavelength (left), and rotational energy levels (right) in the vibrational energy levels.

vibration energy level has an hyperfine structure involving several rotational energy levels due to the rotation of the molecule. Rotation energy levels are identified by a rotation quantum number J. A group of oscillation lines is labeled as $P(J)$, which corresponds to the transition line from upper to lower rotation energy levels with quantum numbers of $J-1$ and J, respectively. Those from $J+1$ to J are labeled by $R(J)$. The P and R branches have many oscillation lines with a wavelength spacing of about 0.02 μm. Their linear gains α_q are usually small; furthermore, the gain saturation coefficient β_q is also small because of the rapid and efficient de-excitation from the lower energy level to the ground state. Therefore, a large steady-state laser power can be obtained based on Eq.(2-90) and a device efficiency higher than 10% can be achieved.

Among the variety of structures in CO_2 lasers, Fig. 3-5(a) represents the most popular structure for a CW laser with a sealed-off discharge tube. The discharge tube is made of glass with a length of about 0.7 ~ 1.5 m and is water-cooled. The inner diameter of the tube is about 10 ~ 20 mm. Although a higher gain is expected for a smaller inner diameter of tube, the diameter cannot be smaller than about 10 mm because it should be larger than the spot size of the lowest transverse mode, which is proportional to the square root of the wavelength (see Eq.(2-3)). The total gas pressure is about 3 ~ 20 Torr with a pressure ratio of about 1 : 4 : 1 (= CO_2 : He : N_2). He gas is used to cool the gas temperature.

One of the lines in the P or R branches can be selected by rotating the surface of the diffraction grating used as one of the cavity mirrors. A single longitudinal mode oscillation can be realized for each line because the longitudinal mode separation is larger than the gain spectral linewidth which is determined by the Doppler effect due to thermal motions of CO_2 molecules.

Figure 3-5(b) shows the structure of a hollow waveguide laser developed to reduce the device's dimensions by confining gases in a capillary discharge tube with an inner diameter of several millimeters. Although the propagation loss through the hollow waveguide is usually large, it can be reduced by using appropriate waveguide materials such as BeO and Al_2O_3 to adjust the value of the complex propagation constant (related to the absorption coefficient and the refractive index). Laser oscillation is achieved if the linear gain due to the gas discharge is larger than the cavity loss which is due mainly to the propagation loss. The focal points of concave mirrors are fixed to the end of the waveguide for wavefront matching at the end of the waveguide. Although this configuration does not satisfy the condition for a stable laser cavity of Eq.(2-1), a stable laser oscillation is realized because the light is confined stably in the hollow waveguide. A steady CW output power, higher than 1 W, is obtained when the cavity length, the total gas pressure and the discharge current are about 25 cm, 70 Torr and 3 mA, respectively.

3.2 Dye Lasers

A dye solution with an organic solvent can be used as a gain medium to obtain laser oscillation. The principal advantage of the dye laser is its wide range of wavelength tuning in the visible region.

Dye lasers can be described by five electronic energy levels and the transitions among these levels. A schematic explanation is given in Fig. 3-7. Each of the electron levels is very wide in energy, broadened by a continuum of vibrational, rotational and solvent energy levels. These levels are divided into two manifolds, the first being a singlet manifold which includes the ground state, the excitation-absorption and the lasing transitions, whereas the second is a triplet manifold of levels connected to the singlet manifold by means of spin-forbidden transitions. Transitions within the triplet manifold constitute losses to either excitation or laser emission. Transitions among these five levels can be divided into those involving either absorption or stimulated emission and those for which nonradiative transition occur either within or between the two manifolds. The singlet levels in this figure

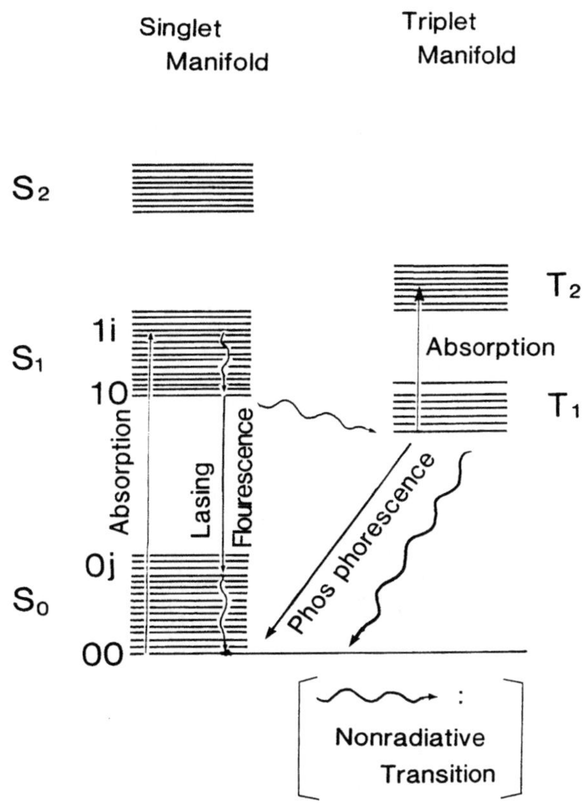

Fig. 3.7. Electron energy levels in a dye molecule [A].

are labeled S_0, S_1 and S_2 for the ground, first excited, and second excited levels, respectively. The lowest and first excited triplet states are correspondingly labeled by T_1 and T_2.

After electrons in S_0 are excited to S_1 by irradiating the pumping light with an appropriate wavelength, the nonradiative transition of electrons to the lowest levels in S_1 occurs within 1 ~ 10 ps and stays at this level for about 1 ~ 10 ns. Subsequently, the transition to S_0 emitting a fluorescence or the nonradiative transition to T_1 follows. The former type of transition is used for the laser oscillation. Population inversion can be realized by optical pumping since de-excited electrons in S_0 immediately fall down to the lowest level in S_0. The lowest level in T_1 is metastable with a long lifetime and the radiative transition to S_0 occurs by emitting phosphorescence. Furthermore, the transition to T_2 can occur by absorption of light. The loss in laser oscillation is increased by this absorption because the absorption wavelength coincides with the fluorescence spectral wavelength for laser oscillation. Therefore, the nonradiative transition from S_1 to T_1 should be avoided. It can be avoided by an appropriate quenching material, called a quencher, which depends on the kind of dye solution.

Over several hundreds of dyes can be used for laser oscillation and the representative fluorescence spectral profile or the gain spectral profile is shown in Fig. 3-8, from which wavelength tunable lasers covering the whole visible range are expected. Among the optical pumping sources for these dyes, Ar^+ or Kr^+ lasers are popularly used because their wavelengths coincide with the absorption wavelength of the transition from S_0 to S_1.

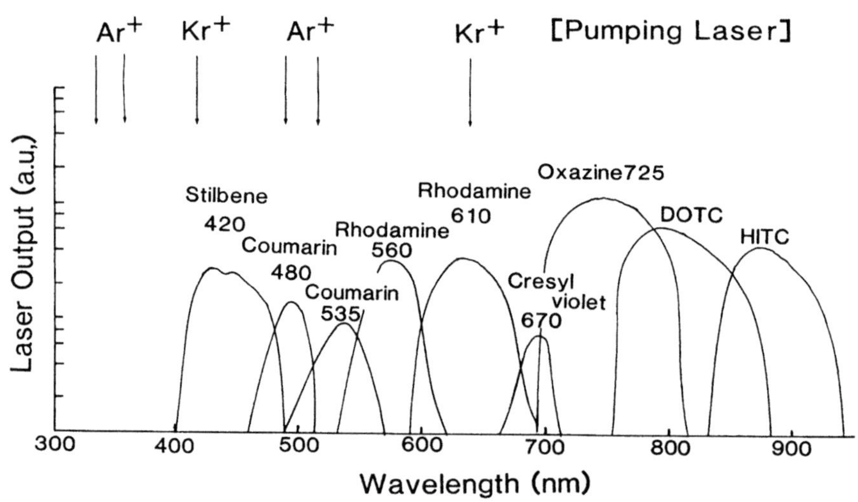

Fig. 3.8. Relation between possible lasing wavelengths of dye lasers and its output power [A].

Figure 3-9 illustrates the structure of a ring cavity dye laser, which is used as a popular CW dye laser. It has been pointed out in Fig. 2-6 that a standing wave can be formed in a conventional Fabry-Perot cavity, in which the linear gain of the lasing mode decreases with an increase in mode power due to spatial hole burning. Since the period of the standing wave which is formed by mode can be different from that of this hole burning, the laser medium still has a linear gain in other modes, by which these modes begin laser oscillation. So the laser operates in a multi longitudinal-mode oscillation with an increase in laser power. However, in Fig. 3-9, spatial hole burning does not appear because the running wave is in an 8-shaped ring cavity made by four mirrors ($M_1 \sim M_4$). A high power and single longitudinal mode oscillation can be realized by eliminating spatial hole burning by this configuration. A thin dye flow film is formed by spouting out from a nozzle at the beam waist of the ring cavity and the pumping light is irradiated through this dye jet. An unidirectional optical element, such as the Faraday rotator, is installed in the cavity to realize a low loss ring cavity only for one of the two running waves. Since the gain spectral linewidth is wide, a birefringence filter, a thin Fabry-Perot etalon and a thick Fabry-Perot etalon are also installed in the cavity in order to achieve a single longitudinal mode oscillation. Figure 3-10 shows the frequency dependences of these optical elements. The longitudinal mode separations of the thin and thick etalons are about 1,000 GHz and 80 GHz, respectively, whereas that of the ring cavity is 200 MHz (for the cavity length of 1.5 m). This figure shows an example of situation in which peak frequencies of the upper three curves coincide with each other while the longitudinal mode frequency of the ring cavity is slightly deviated. Laser oscillation is obtained by the longitudinal mode which is closest to the peak frequencies of the upper three curves. The ring cavity length can be varied by rotating the two quartz plates. The maximum values of the device efficiency and output power are about 30% and 50 W, respectively.

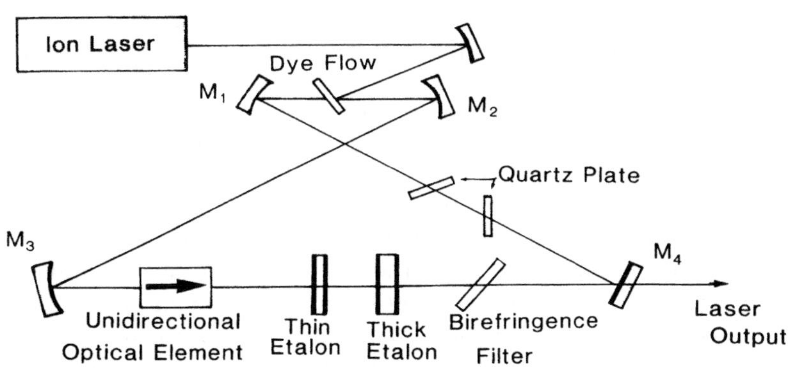

Fig. 3.9. Configuration of a ring cavity-type dye laser [A]. Ring cavity is formed by four cavity mirrors ($M_1 \sim M_4$).

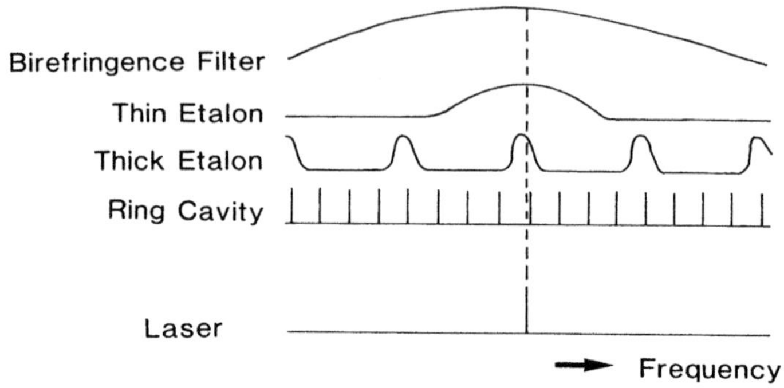

Fig. 3.10. Schematic explanation of dye laser frequency tuning by using different frequency selective devices [A].

3.3 Solid State Lasers

Laser oscillation can be realized by doping impurities such as Neodymium ions (Nd^{3+}) into a host material such as transparent and insulator crystal or amorphous glass in which impurities work as the gain medium. Since the active ion density in crystal is much higher than that of the gas lasers, for example, as high as $10^{17} \sim 10^{20}$ cm^{-3}, a high laser power can be expected. This section reviews the structure and oscillation characteristics of a Nd : YAG laser, several tunable lasers and lasers using a color center produced by irradiating an electron beam onto alkali halide crystals.

A Nd : YAG laser uses a YAG ($Y_3Al_5O_{12}$) crystal in which trivalent rare earth Nd^{3+} ions are doped. Figures 3-11(a) and (b) show the energy levels of Nd^{3+} and laser transitions, where (a) gives the picture for a basic understanding of laser oscillation. Ions are effectively excited from the ground state ① to the level with a band-structure ③ by optical pumping. The excited ions relax within a very short period from level ③ to ② through a nonradiative transition. Level ② which has a long lifetime serves as the upper level for laser oscillation. Level ①', located slightly above level ①, works as a lower level. Sufficient population inversion can be realized between levels ② and ①' because a nonradiative transition from ①' to the closely located level ① easily occurs. As is seen from this figure, a laser utilizing transitions between two of the four energy levels is called a four-level laser. Figure 3-11(b) shows a detailed schematic explanation of energy levels. The energy level of Nd^{3+} is usually splitted into several levels and broadened for a Stark effect induced by an electric field formed by ions around Nd^{3+} in the crystal; moreover, the levels overlap with each other to form an energy band structure. By using the three bands in ①'; laser oscillation at wavelengths of $1.05 \sim 1.12$ μm, $1.3 \sim 1.4$ μm and 1.85 μm can be realized. As an exceptional case, a 0.94 μm wavelength oscillation is obtained by using the level ① as the lower level, which should be called

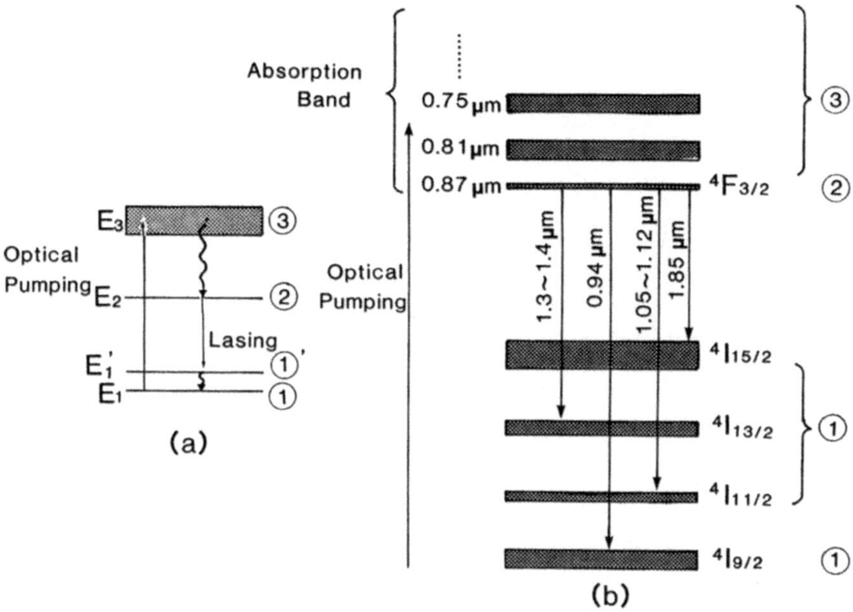

Fig. 3.11. Energy levels of a four-energy level-type solid-state laser [A]. (a) Simplified description. (b) Detailed description to represent wavelengths and energy levels relevant to Nd-YAG laser oscillation.

a quasi-four level system. Among these lines, 1.06 μm is the most useful line because a high laser power is available. There are several absorption bands (level ③) for optical pumping, e.g., wavelengths of 525 nm, 585 nm, 750 nm, 810 nm and 870 nm. Their bandwidth is about 30 nm.

Some types of lamps, such as a Tungsten lamp, Xenon arc lamp and Krypton arc lamp can be used as an optical pumping source. Figure 3-12 shows the structure of a typical CW Nd : YAG laser device, where the laser rod and lamp are installed in parallel on the two focal points of the elliptical cross-section of the reflecting mirror for efficient optical pumping. The cavity structure is almost the same as those for gas and dye lasers. A high quality power source and cooling system are necessary. The device's efficiency is usually as low as 2 ~ 4% and the CW laser power can be as high as 3 ~ 400 W.

Recently, semiconductor lasers have been employed as a pumping source [2] to reduce the power consumption and the device's volume. They can also be used to improve the output beam profile by realizing the lowest order transverse mode, and to realize a direct laser power modulation. AlGaAs lasers at a 810 nm wavelength can be used as efficient pumping sources because their wavelengths coincide with that of the absorption bands shown in Fig. 3-11(b) and high power AlGaAs laser devices are commercially available. Examples of such semiconductor

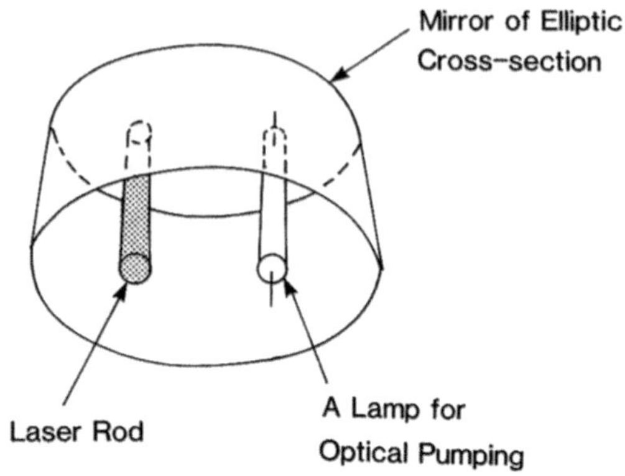

Fig. 3.12. Configuration of a Nd-YAG laser pumped by a lamp [A]. Cavity mirrors are omitted from this figure for simplicity.

laser pumped Nd : YAG lasers are shown in Figs. 3-13(a) and (b), representing a side-pumping and end-pumping scheme, respectively. The structure of the side-pumping device is similar to that of conventional ones (see Fig. 3-12) and an output power of several watts can be obtained by using a semiconductor laser array. For the end-pumping device, a dichromatic cavity mirror should be used and its reflectivity is low and high at wavelengths of 810 nm and 1.06 μm wavelength, respectively. To obtain a high pumping efficiency, a mode matching lens should be used to match the spatial profile of the light beam of the semiconductor laser with that of the Nd : YAG laser in the cavity (see Eq.(2-2)).

Figure 3-14 shows a device which is called a nonplanar ring oscillator (NPRO) or a monolithic isolated single-mode end-pumping ring (MISER) laser, in which the specially polished crystal surface forms a ring cavity and end-pumping is employed [3]. A D.C. magnetic field is applied to the crystal in order to obtain only one running wave by utilizing the Faraday effect or the non-reciprocal polarization rotation effect of the crystal itself. Wavelength can be tuned by varying the crystal temperature. The oscillation is very stable because the cavity is monolithically formed. As another example, a miniature laser using a Nd : YAG microchip with a size of $0.7 \times 1.0 \times 2.0\ mm^3$ and end-pumping has been fabricated [4]. Frequency modulation is possible by applying stress to the crystal from a piezo-electric transducer (the PZT) which is fixed on the side wall of the crystal. The frequency modulation efficiency is about 0.3 MHz per unit voltage applied to the PZT. The highest modulation frequency is as high as 25 MHz and is limited by the resonance frequency of acoustic vibration of the Nd : YAG crystal.

In addition to the Nd : YAG laser described above, it can be expected from Fig. 3-15 that there is a variety of tunable solid state lasers in the near infrared and visible wavelength regions by selecting a combination of ions and host crystals [5]. For example, low noise and all solid state Ti : Al_2O_3 (sapphire) laser are expected within

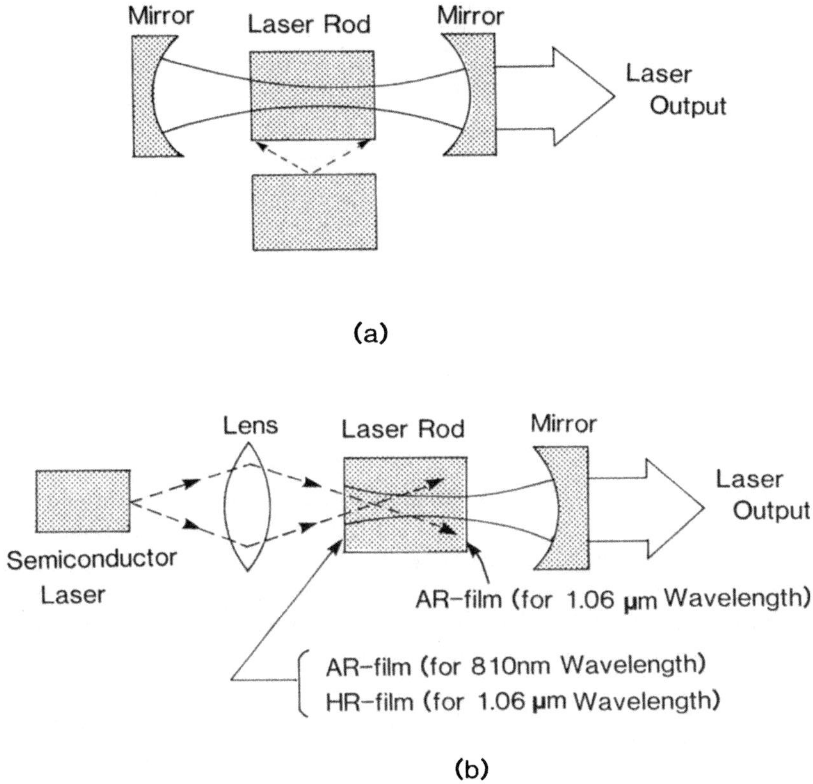

Fig. 3.13. Configuration of a Nd-YAG laser pumped by a semiconductor laser. (a) Side-pump type. (b) End-pump type.

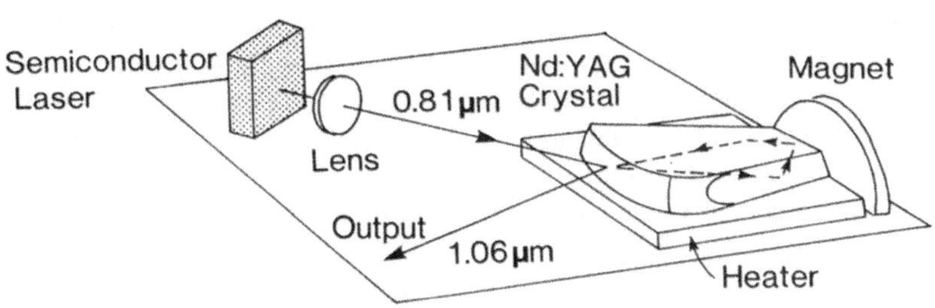

Fig. 3.14. A MPRO-type miniature Nd-YAG lasers pumped by a semiconductor laser. (after Nilsson, Gustafson, and Byer [3], ©1989 IEEE).

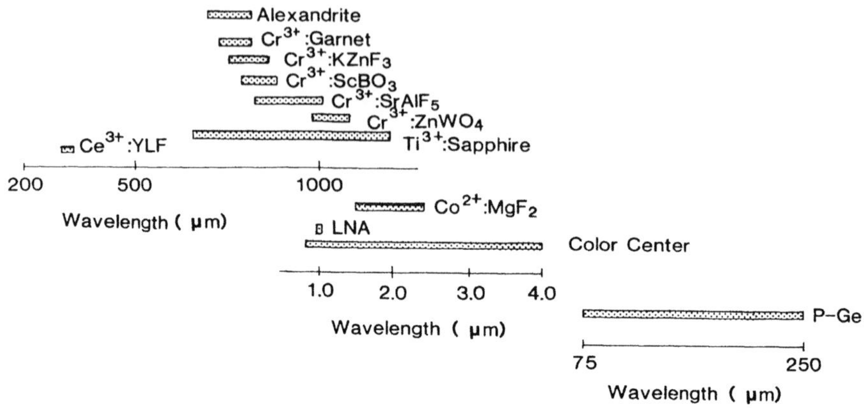

Fig. 3.15. Possible wavelengths of solid-state lasers(after Ishikawa and Imai [5]).

a wavelength range of 0.67 ~ 1.1 μm if pumped by a second harmonic light of a semiconductor laser pumped Nd : YAG laser; consequently, common semiconductor lasers may be replaced by such a tunable laser in a variety of applications [5]. Furthermore, laser oscillation in the 2.1 μm wavelength is possible by replacing Nd^{3+} in the YAG crystal by Holmium ion (Ho^{3+}) which is appropriate for applications to laser radars for eye safety and for its low propagation loss in the atmosphere.

The color center due to the impurity or lattice defect of an alkali halide crystal such as KCl has a wide emission spectral linewidth, which is an advantage in a laser material. Such a property makes the color-center laser similar to the dye laser [6]. The laser-active centers are based on a simple anion (halide ion) vacancy in an alkali halide crystal showing a rock salt structure. This kind of center, called an F-center, consists of a single electron trapped at such a vacancy, where the immediate surroundings are essentially unperturbed. One may easily find that it is similar to a hydrogen atom structure. The energy level structures can be simplified as a four-level system, as in the Nd : YAG laser. The lasing wavelength ranging from 0.89 μm to 3.3 μm is available by using several alkali halide crystals. They have to be usually cooled by liquid nitrogen and high power lasers, e.g., Ar^{+}, dye or Nd : YAG are used as optical pumping sources. Figure 3-16 shows an example of color center laser, in which the cavity structure is similar to that of a dye laser [6].

3.4 Semiconductor Lasers [D]

Although a semiconductor laser can be considered as a kind of common solid state laser, it has unique characteristics due to its small cavity size, large cavity loss and large and fast variations of active electron density in the conduction band. It has been widely used for a variety of industrial applications owing to its small size, low

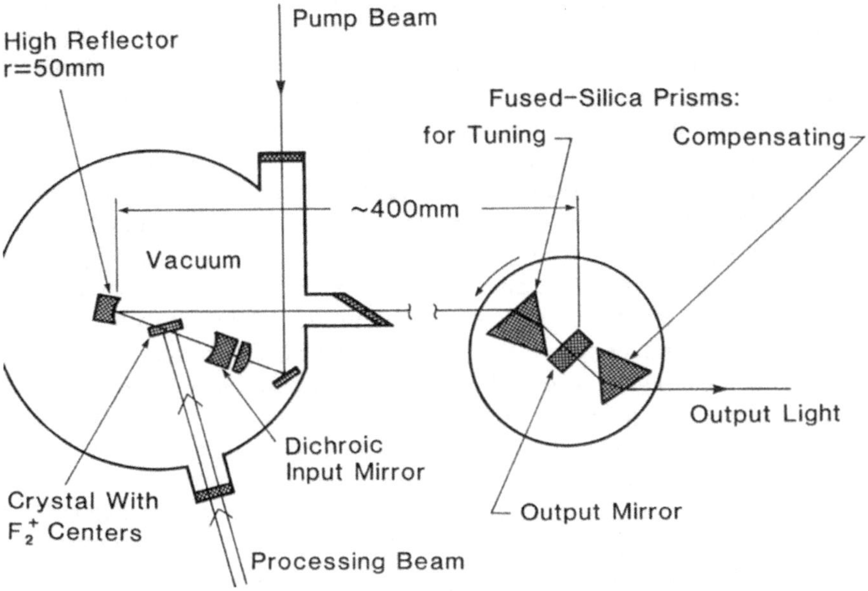

Fig. 3.16. Configuration of a ring cavity-type color center laser (after Mollenauer [6]).

power consumption and low price. Figure 3-17 shows energy band structure of electrons in a semiconductor crystal. The energy band, representing the state of electrons trapped in a crystalline lattice, is called a valence band, whereas the energy band for the state of electrons free from this trap is called a conduction band. These bands are separated from each other and the space between these bands is called the

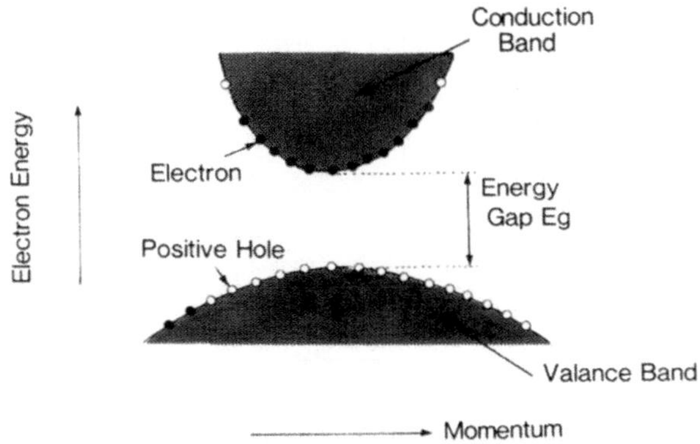

Fig. 3.17. Energy levels of the electrons in a semiconductor crystal.

energy gap. A vacancy is generated after an electron in the valence band has been excited to the conduction band and this corresponds to the appearance of a particle with a positive electric charge which is called positive hole. The electrons in a conduction band or positive holes in the valence band are called carriers.

A direct transition-type semiconductor is used as laser material. In such types of semiconductors, the electron in the conduction band falls down to the valence band, recombines with a positive hole and conserves its momentum without colliding with impurities in the crystal. Due to this transition, spontaneous or stimulated emission occurs. The central frequency of the gain spectral profile in laser oscillation depends on the value of the energy gap, which corresponds to the right-hand side of Eq.(1-47). Direct transition-type semiconductors are found among composite semiconductors which are composed of more than two elements. Since the value of the energy gap can be varied by changing the mole fraction of each element in trinary or quaternary composite semiconductors, lasing wavelengths covering from the infrared to the visible regions can be realized, as shown in Fig. 3-18. The trinary composite semiconductor obtained by doping Al, In or Sb in GaAs has been popularly used in, e.g., compact disk players. It can be expressed as $Al_xGa_{1-x}As$, where x represents the mole fraction ($0 \leq x \leq 1$). The quaternary composite semiconductor $In_xGa_{1-x}As_yP_{1-y}$ has been used in optical fiber communication systems. The development of a visible AlGaInP laser of 0.67 μm wavelength also shows rapid progress. Figure 3-19 shows that the value of the energy gap depends on the mole fractions x and y and that the refractive index decreases with an increase in energy gap; it is thus utilized to confine electrons and light effectively in the semiconductor laser cavity (shown later).

Material		
Active Layer	Clad Layer	Substrate
III-V		
AlGaAs	AlGaAs	GaAs
GaInAsP	GaInP	GaAs
GaInAsP	GaInP	GaAs
GaInAsP	AlGaInP	GaAs
AlGaInP	AlGaInP	GaAs
GaInAsP	AlGaAs	AsAs
GaInAsP	InP	InP
AlGaAsSb	AlGaAsSb	GaSb
InAsSbP	InAsSbP	InAs
IV-VI		
PbSnSeTe	PbSnSeTe	PbTe
II-VI		
ZnSSe		GaAs

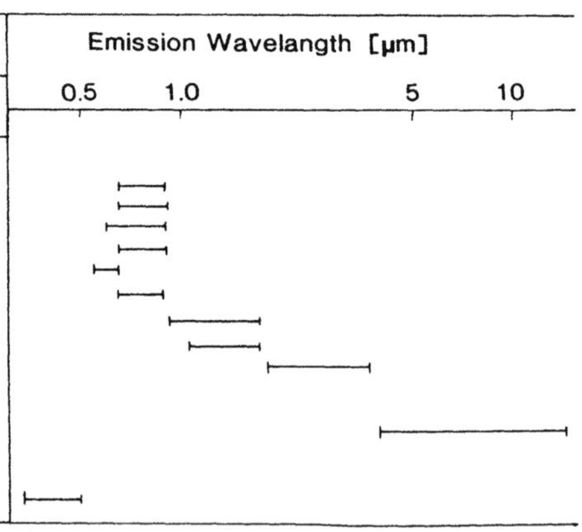

Fig. 3.18. Possible lasing wavelengths by III-V, VI-VI, and II-VI compound semiconductors.

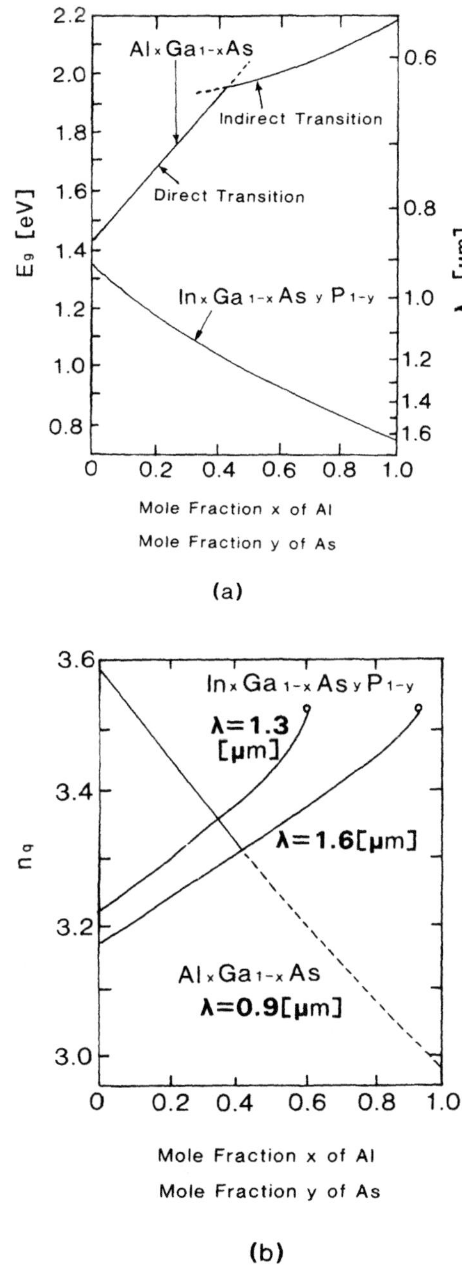

Fig. 3.19. Dependences of energy gap(E_g), corresponding wavelength λ, refractive index n_q on the mole fractions x and y of $Al_xGa_{1-x}As$ and $In_xGa_{1-x}As_yP_{1-y}$ [D]. (a) Dependences of energy gap E_g and wavelength λ (determined by Eq.(1-47)) on x and y. (b) Dependences of refractive index n_q on x and y.

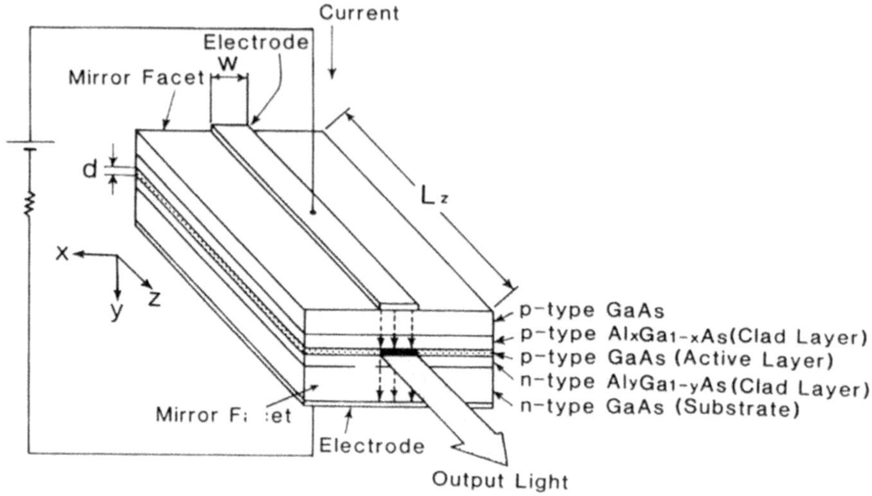
Current
Electrode
W
Mirror Facet
d
Lz
x
y
z
p-type GaAs
p-type AlxGa1-xAs(Clad Layer)
p-type GaAs (Active Layer)
n-type AlyGa1-yAs(Clad Layer)
n-type GaAs (Substrate)
Electrode
Output Light

(a)

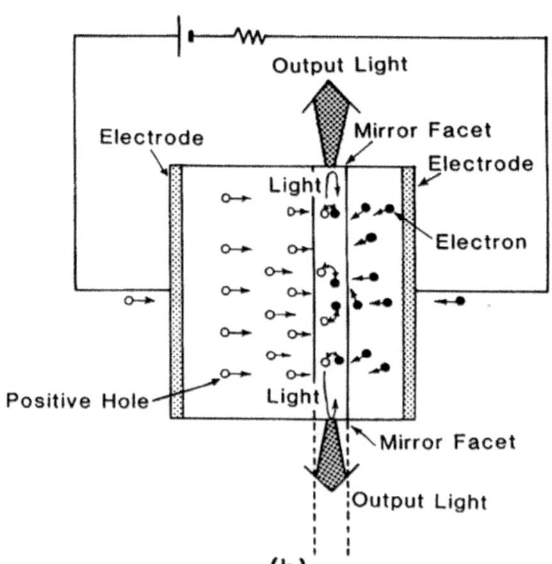
Output Light
Electrode
Mirror Facet
Electrode
Light
Electron
Positive Hole
Light
Mirror Facet
Output Light

(b)

Fig. 3.20

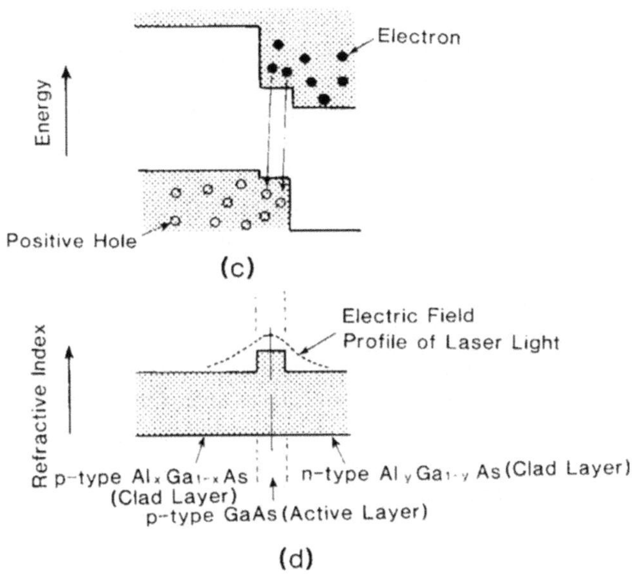

Fig. 3.20. Basic configuration of a semiconductor laser [D]. (a) Illustration of the cavity structure. (b) Cross-section in y-z plane. (c) Energy band. (d) Spatial distributions of refractive index and electric field of laser.

Figures 3-20(a) and (b) show examples of semiconductor lasers and their cross sections on the x-z plane. Light is emitted by the carrier recombination in the GaAs layer of the p-type. Such a p-type layer is called an active layer. The p-type $Al_x Ga_{1-x}As$ and the n-type $Al_yGa_{1-y}As$ layers next to the active layer are called clad layers. A structure like this is called a double hetero-structure because it is composed of two hetero-junctions. After electrons have been injected from a current source into the active layer via the n-type clad layer, they are confined to the active layer because of the energy wall between the active and p-type clad layers, called the hetero barrier (see Fig. 3-20(c)). The spatial distribution of electrons along the y-axis is effectively concentrated in the active layer by the double hetero-structure and the electrons recombine with positive holes while emitting light. Since the diffusion length of the electrons injected into the active layer is about 1~2 μm, the thickness of the active layer is usually designed to be thinner than this value. Furthermore, since the lowest transverse mode oscillation is usually required for application systems, the thickness is fixed to a value as low as approximately 0.1~0.02 μm.

The width of the active layer w along the x-axis is reduced to confine electrons effectively. As was the case for the dielectric waveguide, such as in optical fibers, this reduction simultaneously allows the lowest transverse mode only to oscillate. Although the oscillation condition of the lowest transverse mode is met by $w < 0.5$ μm, the value of w is usually fixed larger than about 2 μm because of the technical limitation of fabricating a narrow stripe electrode.

Now, the electrons are confined in a waveguide cavity having a rectangular cross-section with a size of $d \times w$. The light is confined in the active layer and propagates along the z-axis in the double hetero-structure waveguide, as shown in Fig. 3-20, because the refractive index of the active layer is larger than those of the clad layers, as was shown in Fig. 3-19(b). The typical ratio between the light power confined in the active layer and the total power (this ratio is called the confinement factor) is as low as about 0.2; however, its confinement is sufficient to realize CW and room temperature oscillation.

The cavity length along the z-axis is fixed usually to about 300 μm. The Fabry-Perot cavity is formed by producing flat mirrors on the waveguide facets by cleaving or chemical etching. Lasers having the structure described above are called stripe-type double hetero-structure lasers. The value of the internal loss α_l in Eq.(2-10) is typically about 10 cm^{-1} owing to scattering by the crystal lattice and absorption by impurities. The reflectivity of the cleaved facet is determined by the refractive index of the material n_q, which is given by

$$R_1 = R_2 = \left\{\left(n_q - 1\right)/\left(n_q + 1\right)\right\}^2 \qquad (3-1)$$

The reflectivity of GaAs is 0.3 because $n_q = 3.5$ (dielectric thin films are sometimes coated on the facets to increase or decrease the reflectivity in order to satisfy requirements from practical application systems). As a result, the value of the cavity loss α_{tl} given by Eq.(2-10) is about 50 cm^{-1}. It corresponds to the cavity Q factor of 6×10^3, and implies that the cavity loss is much larger compared to that of other types of lasers. By increasing the injection current I over its threshold value I_{th}, the laser power increases with an increasing current. An example of relationship between the injection current and the laser power, called the *I-L* curve, is given in Fig. 3-21.

Equation (2-8) shows that the adjacent longitudinal mode separation is $\Delta\lambda_q =$ 0.34 nm or $\Delta\nu_q/2\pi = 140$ GHz for $n_q = 3.5$, $\lambda_q = 0.85$ μm, and $L = 300$ μm. The spectral width of the gain is about 100 times larger than the separation, which means that a large number of longitudinal modes can oscillate simultaneously. An example of spectral profile for this multi-longitudinal mode oscillation is shown in Fig. 3-22. Each longitudinal mode also accompanies several transverse modes. To realize the lowest transverse mode oscillation, the technique of transverse mode control is employed at the stage of device fabrication, in which the dimension of the active layer along the x-axis in Fig. 3-20(a) is limited to reduce the size of the cross-section of the active layer, as has been already done for the fabrication of a single mode optical fiber. This can be realized by employing the "index guided" structure, i.e. the difference in refractive index formed along the x-axis to increase the confinement factor for the optical field along this axis. Such a structure has been employed for most commercially available semiconductor lasers. Figure 3-23 shows examples of structures of these lasers [D].

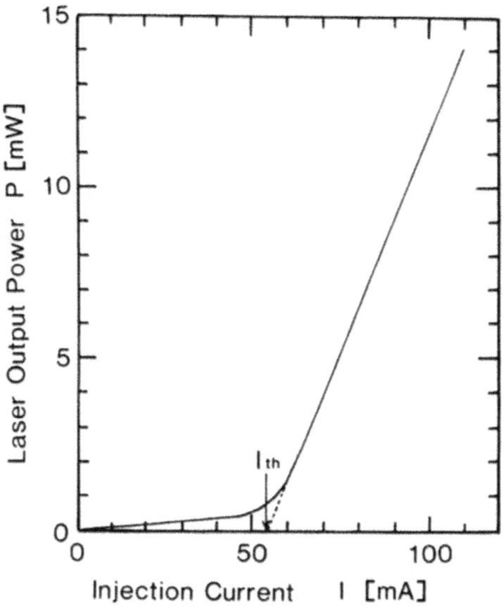

Fig. 3.21. Relation between laser power and injection current (I-L curve) [D].

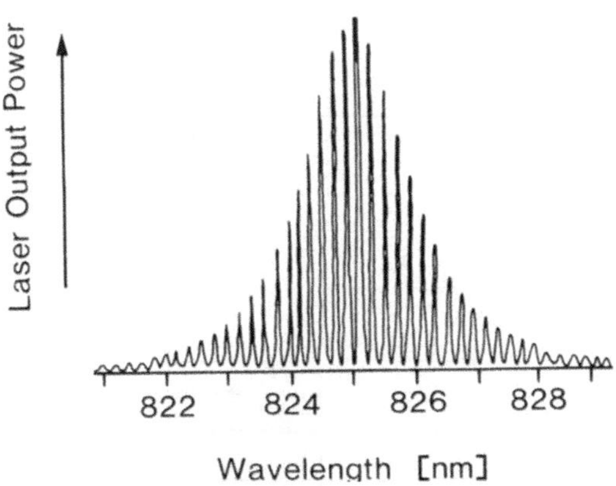

Fig. 3.22. Field spectral profiles of the longitudinal modes of a semiconductor laser oscillation [D].

On the other hand, a laser having a stripe electrode with several μm width is called a gain guided laser, which has its gain within a limited region of the waveguide along the x-axis. The transverse mode control is not effective in this case, and as has been shown in Fig. 3-22, multi-longitudinal mode oscillations are also seen simultaneously. One of the possible reasons for these multi-longitudinal mode oscillations of gain-guided lasers is the appearance of a spatial hole burning.

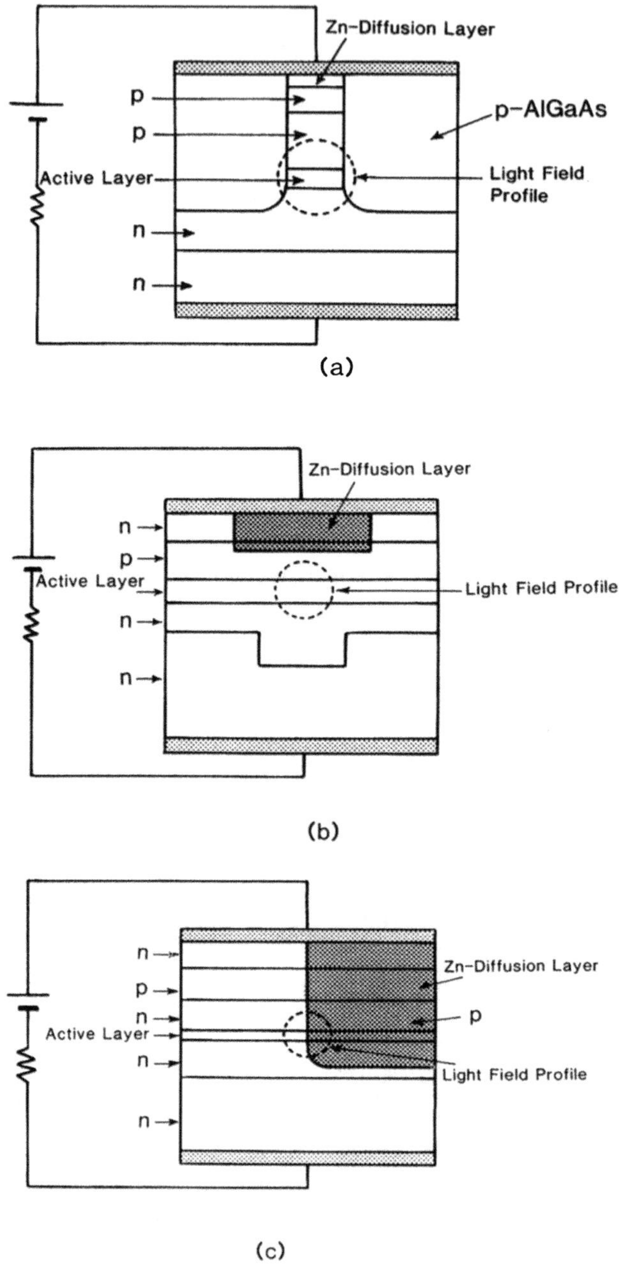

Fig. 3.23. The cross-sectional view of index-guided semiconductor lasers [D]. (a) Buried heterostructure (BH) type. (b) Channeled substrate planar (CSP) type. (c) Transverse junction stripe (TJS) type.

Figure 3-24 shows an example of spectral profile for an index guided laser [D], which shows single-longitudinal mode oscillation. As in that case, it has been empirically found that the transverse mode controlled laser usually shows single longitudinal mode oscillation. However, the jump in laser oscillation from one longitudinal mode to the other occurs when the injection current or temperature changes due to a very small gain difference between adjacent longitudinal modes. This jump is called mode hopping. Furthermore, multi-longitudinal mode oscillations also occur when fast and large amplitude modulations of injection current are applied.

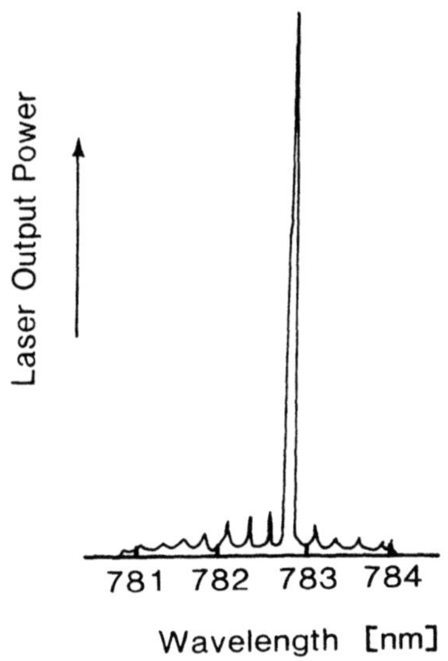

Fig. 3.24. Field spectral profile of the longitudinal modes of an index-guided semiconductor laser.

The technique of longitudinal mode control has been developed by modifying the cavity structure in order to realize a single longitudinal mode oscillation even during fast modulations. For this purpose, a cavity having a low loss only for specific longitudinal modes is used. Figures 3-25(a) and (b) show the typical structures of such lasers, in which diffraction grating is fabricated along the active layer [D]; this type of diffraction grating has been commonly used for other lasers (see Fig. 3-5). Although the resolution of wavelength selection of the diffraction grating is as low as 0.1 nm, it is sufficient for semiconductor lasers because their longitudinal mode separation is as large as about 0.3 ~ 0.5 nm. Figure 3-25(a) represents the laser with a distributed feedback (DFB) cavity, in which a diffraction grating is fabricated on or below the active layer. The wavelength corresponding to the maximum reflectivity of the DFB cavity is

$$\lambda = \lambda_B \pm \delta\lambda_l \qquad (3-2)$$

where λ_B (= $2n_q\ \Lambda/m$) is the Bragg wavelength, Λ is the grating period and m is an integer representing the order of diffraction. $\delta\lambda_l$ is a quantity whose value depends on the groove depth of the grating, the cavity length L and the refractive index. Since the highest reflectivity is realized at two wavelengths which are located at symmetrical positions around the Bragg wavelength, this symmetry is broken by introducing a quarter-wavelength shift of the grating; as a result, a single longitudinal mode oscillation at one of the two wavelengths is realized.

Figure 3-25(b) represents a laser with a distributed Bragg reflector (DBR), in which diffraction gratings are butt-jointed on one or two sides of the active layer.

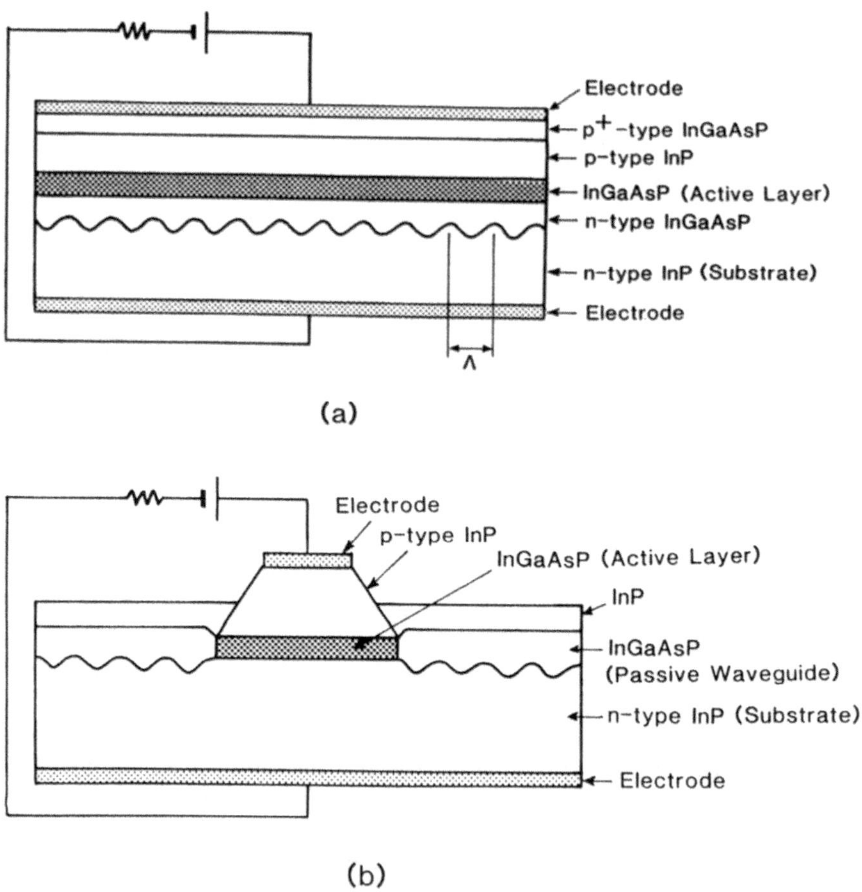

Fig. 3.25. Structures of a longitudinal mode-controlled semiconductor lasers [D]. (a) A distributed feedback (DFB) laser. (b) A distributed Bragg reflector (DBR) laser.

To realize a low absorption loss, diffraction grating is made of the material with a band gap energy larger than that of the active layer material. Although a high accuracy of crystal growth is required to realize the low loss butt-joint, a single longitudinal mode oscillation is always realized because of the lack of symmetry, which is different from the case of (a). However, the output power is lower than that of (a) because of its shorter active layer.

Even in these lasers, the laser power of the other modes becomes larger at a high bias level because of the spatial hole burning effect along the z-axis. Figure 3-26(a) shows a novel device structure developed to reduce this phenomenon by adjusting

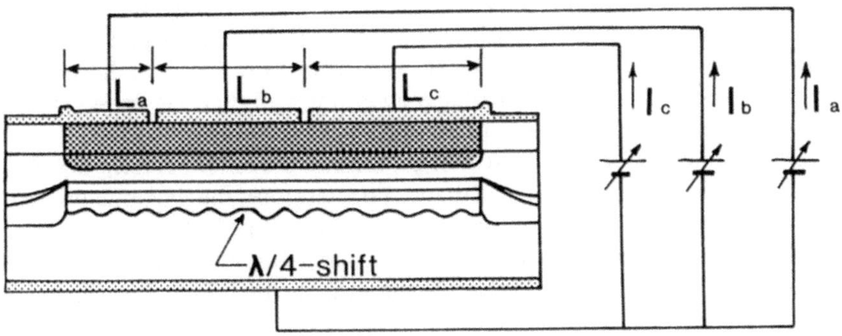

(a)

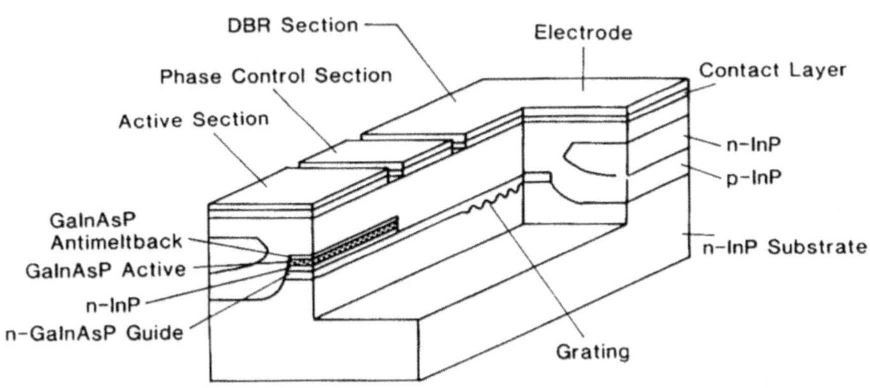

(b)

Fig. 3.26. Structures of semiconductor lasers with segmented electrodes. (a) For depression of spatial hole-burning (after Usami, Akiba and Matsushima [7]). (b) For a wide range wavelength tuning (after Kotaki, H. Ishikawa, Matsuda, and Imai [8]).

the value of the current injected into each segmented electrode [7]. Furthermore, (b) shows a laser with a tuning range over 6.2 nm realized by dividing the electrodes on the active layer and using grating to control the laser power and wavelength [8]. This structure is similar to that of tunable lasers such as the CO_2 laser (see Fig. 3-5). A phase modulator is also fabricated between the two sections to stabilize laser oscillation.

It has been possible to control lasing characteristics, e.g., the gain spectral profile, by changing the energy band structure with the help of the ultra-thin film epitaxial growth technique which has rapidly developed in the past few years. As shown in Figs. 3-27(a) and (b), if the thickness of the active layer is reduced to about 20 nm, which is comparable to the broadening of the electronic wave function, the motion of electrons along the direction perpendicular to the film plane is quantized. The resultant discrete energy levels are shown in Fig. 3-27(b) which represents the quantum-size effect appeared in the gain and the loss properties of the material [C]. A laser employing this effect is called a quantum well (QW) laser. As is shown in Fig. 3-28, since the gain spectral linewidth of such a laser is narrower than that of conventional lasers for quantized energy, the low threshold current density, high device efficiency and so on, are realized. Furthermore, the dynamic property and noise characteristics are also improved. The thin film exhibiting a quantum-size effect has already been developed and new lasers with two- and three-dimensional quantum-size effects are also under development; they are called quantum wire

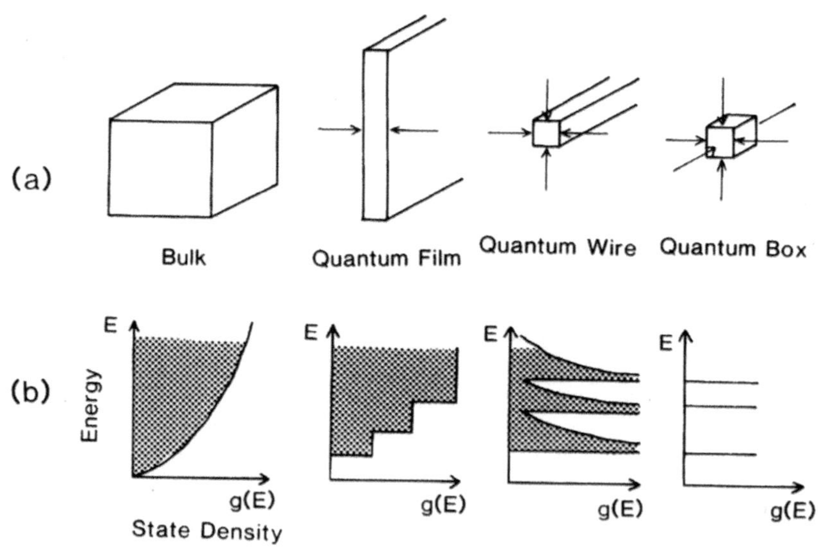

Fig. 3.27. Semiconductor laser structures with (a) a micro-sized active region (b), and the profiles of their energy bands [C].

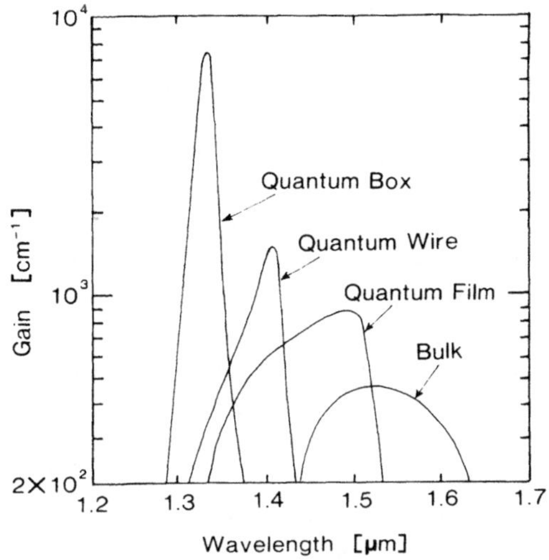

Fig. 3.28. Gain spectral profiles of semiconductor lasers with a micro-sized active region [C].

lasers and quantum box lasers, respectively. Although gain is increased and gain spectral linewidth is decreased by enhanced confinements, these lasers still show a multi-longitudinal mode oscillation because the linewidth is still larger than the separation of the longitudinal mode. The technique of longitudinal mode control is still required for these lasers.

The threshold current depends on the cavity loss and the width of the active layer. The typical value ranges from 1 mA to several 10 mA at room temperature. Quantum well lasers demonstrate a very low threshold of about 0.55 mA [9]. The practically usable output power ranges from several mW to several 10 mW at room temperature. Power as high as several 100 mW has been realized by gain-guided and multi-longitudinal mode lasers, e.g., 300 mW output power was obtained by introducing a broad stripe structure into a 0.67 μm wavelength visible AlGaInP laser [10]. Total output power, as high as 76 W, was also realized by an AlGaAs laser array, which is composed of twenty closely located active waveguides fabricated in parallel on a common substrate [11].

The device efficiency, defined by the ratio between light output power and electrical power supplied from the current source, is about 10%, which is comparable to that of a CO_2 laser. However, the differential quantum efficiency, defined as the increase of photon number per unit increase of electrons injected from the current source, is as high as 40–60%. This efficiency corresponds to the value of the slope of the *I-L* curve in Fig. 3-21. Due to such a large differential quantum efficiency, semiconductor lasers are regarded as highly efficient lasers.

The divergence angle of the laser beam emitted from the cavity is very large due to large diffraction effect under the condition of $d < \lambda < w$, where d, w and λ

represent the thickness, width of the active layer and wavelength, respectively. In addition, the divergence angle along the direction of the active layer surface θ_w is smaller than the angle θ_d along the direction perpendicular to it for the relation given above. Their typical values are about 15 and 30 degrees, respectively. Equation (2-5) shows that such a divergence angle is much larger than that in other types of lasers and that the output laser beam does not have a circular cross-section. A microscope objective lens is required to collimate the output laser beam.

Although the carrier density in the conduction band is clipped to its threshold value (see Eq.(A-13) in Appendix) when a d.c. current is injected into the laser, the carrier density is modulated by modulating the injection current, which can modulate the laser power. Since it is realized by modulating the carrier density directly in the active medium, it is possible to realize a fast and high efficiency modulation. This is the specific property of semiconductor lasers, being different from the modulation schemes of other types of lasers. Figure 3-29 shows the schematic explanation of laser power modulation characteristics (the intensity modulation: IM) [12]. The frequency f_r in this figure represents a relaxation oscillation frequency (see Eq.(A-21) in Appendix). The factor 2 in the last term on the right-hand side of Eq.(A-11) (the rate equation for the carrier density) can be removed because one pair of electron and positive hole disappears after emitting a photon, (i.e. an active electron in the conduction band is trapped in the valence band). Therefore, the factor 2 is also removed from Eq.(A-21), which gives the following expression for f_r:

$$f_r = \frac{1}{2\pi\sqrt{\tau_p \tau_s}} \sqrt{I / I_{\rm th} - 1} \qquad (3-3)$$

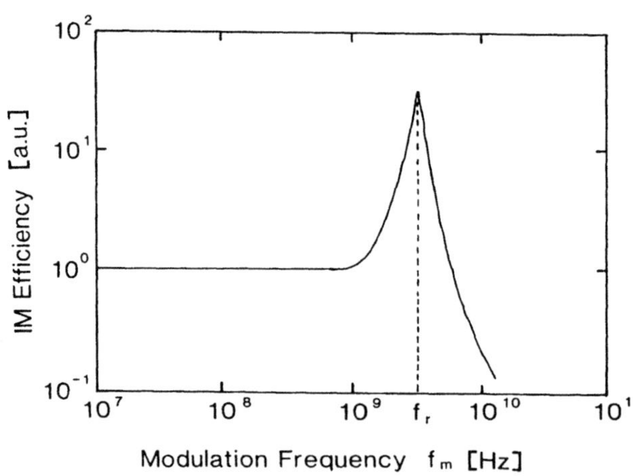

Fig. 3.29. Direct intensity modulation characteristics of a semiconductor laser [D].

where I, I_{th}, τ_p and τ_s are the D.C. injection current, the threshold value, the photon lifetime (Eq.(2-17)) and the inverse of the longitudinal relaxation constant (see Appendix), respectively. The value of f_r is as large as several GHz because the typical values of τ_p and τ_s are about 1 ps and 3 ns, respectively. A novel laser device with f_r as high as several 10 GHz is also possible by employing a quantum well structure.

The laser frequency can also be modulated directly by the injection current because the longitudinal mode frequency is modulated by changes in the refractive index n_q and the cavity length L of Eq.(2-7). A change in the frequency is expressed as

$$\Delta \nu_l = -\nu_l \left[\left(\frac{A}{n_q} \right) \Delta N_c(I) + (\alpha_T + \beta_T) \Delta T(I) \right] \tag{3-4}$$

The first-term on the right-hand side represents change in the refractive index n_q due to the carrier density change $\Delta N_c(I)$ induced by change in injection current. The quantity A is a proportional constant. The second- and third terms represent the thermal effect of the cavity length L (its thermal expansion coefficient is given by $\alpha_T \equiv \partial L/\partial T$) and change in the refractive index n_q (its thermal coefficient is given by $\beta_T = \partial n_q/\partial T$), respectively. Both effects result from temperature change $\Delta T(I)$ in self-heating of semiconductor materials caused by a change in injection current. Typical values for these parameters are given in Table 3-1 [D]. Figure 3-30 shows the frequency modulation (FM) characteristics of several lasers which are represented by the transfer function H_L ($\equiv \partial \nu(f_m)/\partial I(f_m)$) [13]. The transfer function H_L usually takes a complex value because the modulated laser frequency $\nu(f_m)$ usually has a phase delay with respect to the phase of the modulation current with the modulation frequency f_m. Figures 3-30(a) and (b) represent the absolute value and phase angle of H_L. When the modulation frequency $f_m <$ several MHz, the thermal effect is dominant for FM because the second- and third terms of Eq.(3-4) are larger than the first term. The carrier effect becomes dominant for $f_m >$ several MHz for the self-heating and the temperature change of the semiconductor material cannot follow the fast change in modulation current. The carrier effect has a cutoff

Table 3.1. Values of coefficients in Eq.(3-4) [D].

	AlGaAs Laser (0.8 μm)	InGaAsP Laser (1.5 μm)
A	-4.0×10^{-27} m^3	-7.0×10^{-27} m^3
n_q	3.5	3.5
α_T	5×10^{-6} K^{-1}	5.4×10^{-6} K^{-1}
β_T	1×10^{-4} K^{-1}	1.0×10^{-4} K^{-1}

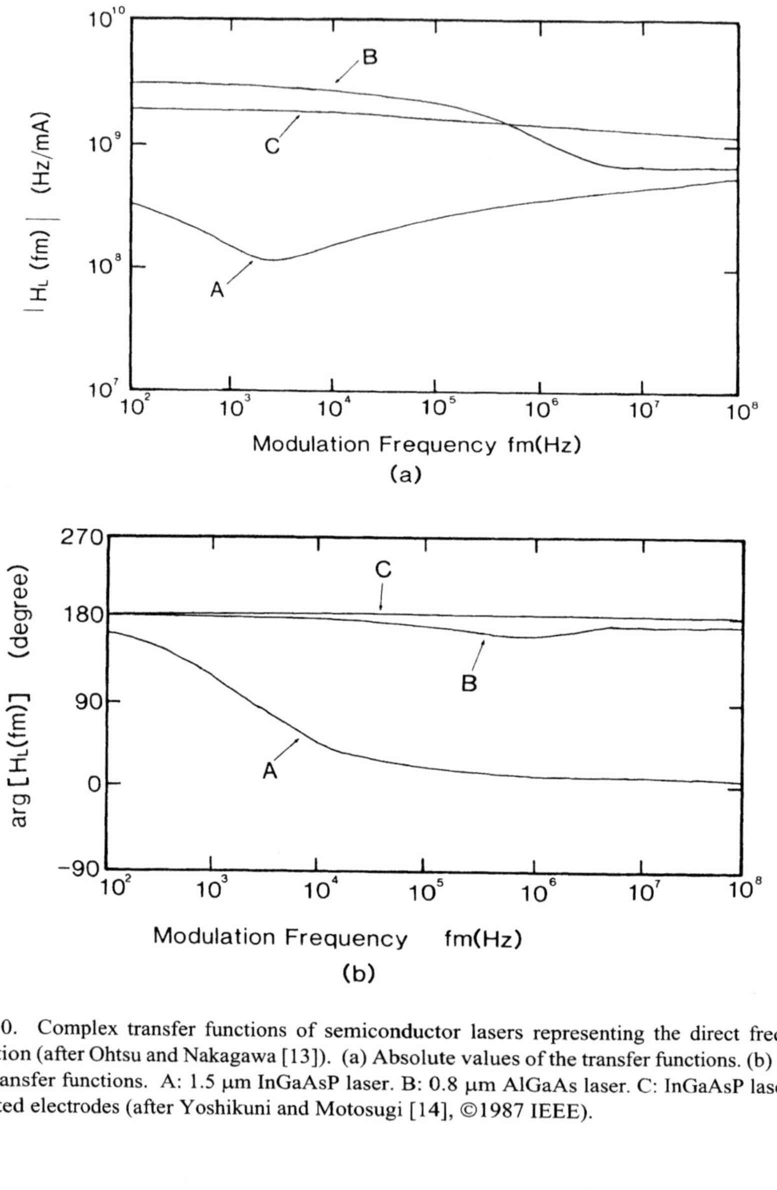

Fig. 3.30. Complex transfer functions of semiconductor lasers representing the direct frequency modulation (after Ohtsu and Nakagawa [13]). (a) Absolute values of the transfer functions. (b) Phases of the transfer functions. A: 1.5 μm InGaAsP laser. B: 0.8 μm AlGaAs laser. C: InGaAsP laser with segmented electrodes (after Yoshikuni and Motosugi [14], ©1987 IEEE).

frequency determined by the relaxation oscillation frequency f_r (see Eq.(3-3)). In each of the curves in these figures, there exists a turning point at the f_m of several MHz and the contributions from the thermal and carrier effects cross at this frequency. The profiles of these curves depend on the waveguide structure, semiconductor materials and so on. A flat profile, advantageous for practical applications, has been realized by inhomogeneous distribution of carriers for which a laser with segmented electrodes has been developed to control the currents injected into these electrodes. An example is shown in Fig. 3-30 [14].

[Textbooks]

[A] T. Tako and M. Ohi, *Introduction to Lasers,* Kyoritsu Publishing, Tokyo, 1985 (in Japanese).

[B] C. Yamanaka (Chief Editor), *Laser Handbook,* Ohm-sha Publishing, Tokyo, 1978 (in Japanese).

[C] Y. Yajima, K. Shimoda, F. Inaba and S. Nanba (Editors), *Laser Handbook,* Asakura Publishing, Tokyo, 1989 (in Japanese).

[D] M. Ohtsu, "Technology for Light Emitting and Detecting Devices for Fiber Sensor", Chapter 3 in *Optical Fiber Sensor,* ed. by T. Okoshi, Ohm-sha Publishing, Tokyo, 1986 (in Japanese).

[References]

1) W. B. Bridges, "Atomic and Ionic Gas Lasers", in *Quantum Electronics*, Part A, ed. by C. L. Tang (Method of Experimental Physics, Vol. 15), Academic Press, New York, 1979.

2) H. Fukube and Y. Ohtake, "Diode laser pumped solid-state lasers", *Optronics*, No. 78 (1988) pp. 87–93 (in Japanese).

3) A. C. Nilsson, E. K. Gustafson and R. L. Byer, "Eigenpolarization Theory of Monolithic Nonplanar Ring Oscillators", *IEEE J. Quantum Electron.*, Vol. 25, (1989) pp. 767–790.

4) J. J. Zayhowski and A. Mooradian, "Frequency-modulated Nd : YAG microchip lasers", *Opt. Lett.*, Vol. 14, (1989) pp. 618–620.

5) K. Ishikawa and S. Imai, "Wavelength tunable solid state lasers", *Optronics*, Vol. 78, (1988) pp. 81–86 (in Japanese).

6) L. F. Mollenauer, "Color Center Laser", in *Quantum Electronics*, Part B, ed. by C. L. Tang (Method of Experimental Physics, Vol. 15), Academic Press, New York, 1979.

7) M. Usami, S. Akiba and Y. Matsushima, "Mode characteristics of λ/4-shifted DFB lasers with the distributed current injection along the cavity", Extended Abstracts (The 49th Autumn Meeting, 1988); The Jpn. Soc. Appl. Phys., Oct., 1988, 6pZC16 (in Japanese).

8) Y. Kotaki, M. Matsuda, H. Ishikawa and H. Imai, "Tunable DBR laser with wide tuning range", *Electron. Lett.*, Vol. 24, (1988) pp. 503–505.

9) K. Y. Lau, P. L. Derry and A. Yariv, "Ultimate limit in low threshold quantum well GaAlAs semiconductor lasers", *Appl. Phys. Lett.*, Vol. 52, (1988) pp. 88–90.

10) K. Itaya, G. Hatakoshi, M. Ishikawa, Y. Watanabe, and Y. Uematsu, "High power operation of InGaAlP laser diode by broad stripe structure", Extended Abstracts (The 50th Autumn Meeting, 1989), The Japan Society of Appl. Phys., September, 1989, 28aZG1 (in Japanese).

11) M. Sakamoto, D. F. Welch, J. G. Endriz, D. R. Scifres and W. Streifer, "76 W cw monolithic laser diode arrays", *Appl. Phys. Lett.*, Vol. 54, (1989) pp. 2299–2300.

12) Y. Yamamoto, S. Saito, and M. Mukai, "AM and FM quantum noise in semiconductor lasers-Part II: Comparison of theoretical and experimental results for AlGaAs lasers", *IEEE J. Quantum Electron.*, Vol. QE-19, (1983) pp. 47–53.

13) M. Ohtsu and K. Nakagawa, "Frequency Control of Semiconductor Lasers and Its Applications", *Oyo Buturi* (*J. Jpn. Soc. Appl. Phys.*), Vol. 58, (1989) pp. 1428–1444 (in Japanese).

14) Y. Yoshikuni and G. Motosugi, "Multielectrode Distributed Feedback Lasers for Pure Frequency Modulation and Chirping Suppressed Amplitude Modulation", *IEEE J. Lightwave Technol.*, Vol. LT-5, (1987) pp. 516–522.

Chapter 4

Noises in Lasers

Noise properties in laser devices are discussed in this chapter. The fundamental noises are intrinsic quantum noises due to the fundamental mechanisms of laser operation. In addition to these, external noises come from power sources, environmental fluctuations, and so on. The magnitude of external noises is usually larger than from intrinsic noises. Conventional measures used to represent the statistical properties of these noises are defined in 4.1 and discussions on noises are described in 4.2 and 4.3.

4.1 Definitions of the Statistical Measures of Noise Properties and Measurement Procedures [A, B]

Statistical measures to represent the properties of noises are defined in this section. For the randomly fluctuating time-dependent physical quantity $y(t)$, the power spectral density $S_y(f)$ can be used as the first statistical measure to represent the property of $y(t)$. The auto-correlation function defined in the following equation should be calculated to derive $S_y(f)$:

$$R_y(\tau) \equiv \langle y(t')y(t'+\tau)\rangle \equiv \lim_{T\to\infty}\frac{1}{T}\int_{-T/2}^{T/2} y(t')y(t'+\tau)\mathrm{d}t'. \qquad (4-1)$$

This function represents the correlation between $y(t')$ and $y(t'+\tau)$. Based on this function, the power spectral density $S_y(f)$ is defined as the Fourier transform of $R_y(\tau)$, which is

$$S_y(f) = 4\int_0^{\infty} R_y(\tau)\cos(2\pi f\tau)\mathrm{d}t. \qquad (4-2)$$

This expression is called a Wiener-Kinchine's formula. The power spectral density $S_y(f)$, defined in the Fourier frequency range of $0 \leq f < +\infty$, is called a one-sided power spectral density. For theoretical discussion, a two-sided power spectral density has also been popularly used and is defined in the Fourier frequency range of $-\infty < f < \infty$. However, since the negative frequency does not have a physical sense, the one-sided power spectral density is more practical for experimental studies. For reference, the two-sided power spectral density $S_{yt}(f)$ is given by

$$S_{yt}(f) = \frac{1}{2} S_y(f). \qquad (4-3)$$

The auto-correlation function $R_y(\tau)$ is derived by the inverse Fourier transform, i.e.

$$R_y(\tau) = \int_0^{\infty} S_y(f) \cos(2\pi f \tau) df. \qquad (4-4)$$

By substituting $\tau = 0$ into Eqs.(4-1) and (4-4), one has

$$\left\langle \{y(t')\}^2 \right\rangle = \lim_{T \to \infty} \frac{1}{T} \int_{-T/2}^{T/2} \{y(t')\}^2 dt = \int_0^{\infty} S_y(f) df. \qquad (4-5)$$

If one assumes that $y(t)$ represents the voltage in an electronic circuit, the left-hand side of Eq.(4-5) is proportional to the time average of the electric power. Therefore, $S_y(f)$ is called the power spectral density because it corresponds to the density of the power in the unit frequency bandwidth around f. If the unit of $y(t)$ is represented by $[y]$, it is easily seen from Eqs.(4-1) and (4-2) that the dimension of $S_y(f)$ is $[y]^2$/Hz. For example, if the power spectral density is proportional to f^0, f^{-1} and f^{-2}, these noises are called white noise, flicker noise and random walk, respectively.

It is advantageous to use a second measure defined in the time domain in order to evaluate very slow fluctuations quantitatively. To define such a measure, the time-averaged value of $y(t)$ in the period τ is derived as

$$\bar{y}_k \equiv \frac{1}{\tau} \int_{t_k}^{t_k + \tau} y(t) dt, \qquad (4-6)$$

where τ is called the integration time, $k = 0, 1, 2, 3, ..., N$, and $t_{k+1} = t_k + T$. T represents the period of successive measurements, which means that these measurements have a dead time of $T-\tau$. If this deadtime is negligibly small, i.e. $T \cong \tau$, the second measure in the time domain can be defined as

$$\sigma_y^2(\tau) \equiv \left\langle \frac{(\bar{y}_{k+1} - \bar{y}_k)^2}{2} \right\rangle$$

$$= \lim_{N\to\infty} \frac{1}{N-1} \sum_{k=1}^{N} \frac{\left(\bar{y}_{k+1} - \bar{y}_k\right)^2}{2} \qquad (4-7)$$

which is called the Allan variance [1]. It is the most fundamental measure in the time domain and has several properties, as listed below:

(1) Since $\sigma_y^2(\tau)$ is given by the difference between two measured values, $\bar{y}_{k+1}$ and $\bar{y}_k$, it converges even when the value of $S_y(f)$ diverges at $f \to 0$. Therefore, it is advantageous to estimate the magnitudes of flicker noise and random walk.

(2) Since $\sigma_y^2(\tau)$ is given by the sum of $\left(\bar{y}_{k+1} - \bar{y}_k\right)^2$, it is invariant even if $\bar{y}_{k+1}$ and $\bar{y}_k$ are exchanged in Eq.(4-7). It means that $\sigma_y^2(\tau)$ could loose some information on the time sequence contained in $y(t)$.

(3) If the available electronic measurement instruments are used, the measurement accuracy of $\sigma_y^2(\tau)$ is usually higher than that of $S_y(f)$.

A conversion formula, from $S_y(f)$ to $\sigma_y^2(\tau)$, is given by

$$\sigma_y^2(\tau) = 2\int_0^{\infty} S_y(f) \frac{\sin^2(\pi f\tau)}{(\pi f\tau)^2} \sin^2(\pi f\tau) \mathrm{d}f \qquad (4-8)$$

The function $\sin^2(\pi f\tau)/(\pi f\tau)^2$ in this equation represents the rectangular time window with a height of $1/\tau$ and a width of τ, through which the signal $y(t)$ passes, i.e. the value of $y(t)$ is measured for the time period τ. Furthermore, the function $\sin^2(\pi f\tau)$ is the window function representing the measurement without any dead time ($T = \tau$). This window has the characteristics of a high pass filter, i.e. it approaches 0 at the limit of $\pi f\tau \to 0$. Due to such a feature, the value of $\sigma_y^2(\tau)$ does not diverge even in cases of flicker noise and random walk.

The characteristics of angular frequency fluctuations of self-sustained oscillators, including lasers, are discussed in the following. The electric field phase of laser light at time t is expressed as

$$\Phi(t) = \nu_{ql} t + \phi(t) \qquad (4-9)$$

where ν_{ql} is the average angular frequency and $\phi(t)$ is the fluctuating phase. Thus, the instantaneous angular frequency is given by

$$\nu_{ql}(t) = \frac{\mathrm{d}\Phi(t)}{\mathrm{d}t} = \nu_{ql} + \frac{\mathrm{d}\phi(t)}{\mathrm{d}t} \qquad (4-10)$$

where the value of the second term is assumed to be far smaller than that of the first term. The normalized angular frequency fluctuations are given by

$$y(t) = (\mathrm{d}\phi / \mathrm{d}t) / \nu_{ql} \tag{4-11}$$

The power spectral density of $y(t)$ for stable oscillators and single mode lasers can be approximated as

$$S_y(f) \propto \sum_{n=-2}^{2} h_n f^n \tag{4-12}$$

where h_n is the expansion coefficient ($-2 \leq n \leq 2$). For each term on the right-hand side of this equation, an expression corresponding to $\sigma_y^2(\tau)$ is derived from Eq.(4-8) and is summarized in Table 4-1 [A, B], where it is assumed that a rectangular low pass filter with a pass band of $0 \leq f \leq f_h$ is used to measure the value of $S_y(f)$ in order to avoid divergence for $n \geq 1$. The power spectral density of the phase fluctuations $S_\phi(f)$ is derived from $S_y(f)$ in Eqs.(4-10) and (4-11) and is expressed as

$$S_\phi(f) = \left(\frac{\nu_{ql}}{2\pi f} \right)^2 S_y(f) \tag{4-13}$$

It is not straightforward to derive a general conversion formula from $\sigma_y^2(\tau)$ to $S_y(f)$. This is because the information on the time sequence has already been lost from $\sigma_y^2(\tau)$, as pointed out in (2) of the properties listed above. It can be concluded that the power spectral density is a more basic and important measure than $\sigma_y^2(\tau)$. However, in most practical systems, mutual conversion can be carried out by following Table 4-1.

The two measures defined above have been used to represent the magnitude of FM noise, i.e. the angular frequency stability. However, to evaluate laser performances quantitatively, it is insufficient to discuss stability only. Two additional performances, i.e. reproducibility and accuracy, should also be evaluated. The physical significance of these three measures is summarized as:

(1) Stability: A sequence of N readings of a particular oscillator frequency in a particular adjustment against a reference oscillator assumed temporally constant will show fluctuations. The standard deviation of these observations is called the stability [2].

(2) Reproducibility: A sequence of comparisons for the independent adjustments of a particular oscillator frequency against an available and temporally constant reference oscillator frequency will yield a standard deviation. The standard deviation of such observations may be called the reproducibility. In this sense, reproducibility indicates the degree of resettability of the frequency [2].

Table 4.1. Relation between one-sided power spectral density $S_y(f)$ and the Allan variance $\sigma_y^2(\tau)$ [A, B].

One-sided Power Spectral Density $S_y(f)$	Allan Variance $\sigma_y^2(\tau)$
$h_2 f^2$ (*) $(2\pi f_h \tau \gg 1)$	$h_2 \frac{f_h}{(2\pi)^2} \tau^{-2}$
$h_1 f^1$ (*) $(2\pi f_h \tau \gg 1)$	$h_1 \frac{1}{(2\pi)^2}\left(\frac{9}{2} - 3\ln(2\pi f_h \tau) - \ln 2\right) \tau^{-2}$
$h_0 f^0$	$\frac{1}{2} h_0 \tau^{-1}$
$h_{-1} f^{-1}$	$(2\ln 2) h_{-1} \tau^0$
$h_{-2} f^{-2}$	$\frac{(2\pi)^2}{6} h_{-2} \tau^1$

(*) Measured by using a low-pass filter with the cut-off frequency f_h

(3) Accuracy: Accuracy implies fractional uncertainty in determining the atomic transition frequency of free atoms and is expected within 3σ limits for statistically determined frequencies [3].

Frequency fluctuation (FM noise) characteristics can also be evaluated by using the linewidth of a field spectral profile of laser oscillation. The field spectral profile can be derived from Eqs.(4-1) and (4-9) because it corresponds to the power spectral density of a laser electric field. The auto-correlation function of the electric field is expressed as

$$R_E(\tau) = \langle E(t')E^*(t'+\tau)\rangle$$

$$= |E_0|^2 \langle \exp[-i\nu_{ql}\tau - i\{\phi(t'+\tau) - \phi(t')\}]\rangle, \qquad (4-14)$$

where E_0 is the amplitude of the electric field. By following Rowe's theorem in communication theory [4], Eq.(4-14) is transformed by using the two-sided power spectral density $S_{\nu t}(f)$ of the angular frequency fluctuations to

$$R_E(\tau) = |E_0|^2 \exp\left[-i\nu_{ql}\tau - \frac{\tau}{\pi}\int_0^\infty S_{\nu t}(f)\frac{\sin^2(\pi f\tau)}{(\pi f\tau)^2} d(\pi f\tau)\right]. \qquad (4-15)$$

Under the assumption of white noise

$$S_{\upsilon t}(f) = a(= \text{constant}) \qquad (4-16)$$

Eq.(4-15) is transformed to

$$R_E(\tau) = |E_0|^2 \exp\left(-i\nu_{ql}\tau - \frac{a\tau}{2}\right) \qquad (4-17)$$

Then, the field spectral profile is given by the absolute value of the Fourier transform of $R_E(\tau)$, which is proportional to the Lorentzian given by

$$I(f - \nu_{ql}) \propto \frac{1}{(f - \nu_{ql}/2\pi)^2 + (a/4\pi)^2} \qquad (4-18)$$

Its FWHM (full width at half maximum), i.e. its half linewidth Δf, is, therefore,

$$\Delta f = a/2\pi \ (Hz) \qquad (4-19)$$

which is proportional to the magnitude of white noise. The half linewidth Δf has been used as a convenient measure to represent the magnitude of FM noise. However, it is not always an accurate measure when FM noise can not statistically be expressed by white noise.

Figure 4-1 shows the setup to measure FM noise magnitudes. (a) represents the setup to obtain a beat signal between two independent lasers (their average angular frequencies are ν_1 and ν_2) which are incident onto a square-law photo-detector, i.e. the output current is proportional to the square of the electric field amplitude of the incident lights. If the FM noise statistics of the two lasers are uncorrelated between each other within the range of Fourier frequency f or integration time τ under test, the superposition of their field spectral profiles can be displayed by applying an output signal from the photo-detector to the RF spectrum analyzer. The sum of the power spectral density of the two lasers' FM noises can be displayed if a frequency-to-voltage converter (F/V converter) is connected between the photo-detector and the RF spectrum analyzer. The Allan variance can be calculated by applying an output signal from the photo-detector to the frequency counter. The Allan variance real-time processing system has been developed for this purpose [5]. If the magnitude of FM noise of one of the lasers is far larger than that of the other, the results obtained by the setups in Fig. 4-1 represent the magnitude of the FM noise of the former since the latter works as a frequency reference.

(b) and (c) represent simpler methods of using a high precision scanning Fabry-Perot interferometer and a fiber delay line [6], respectively. Although no reference

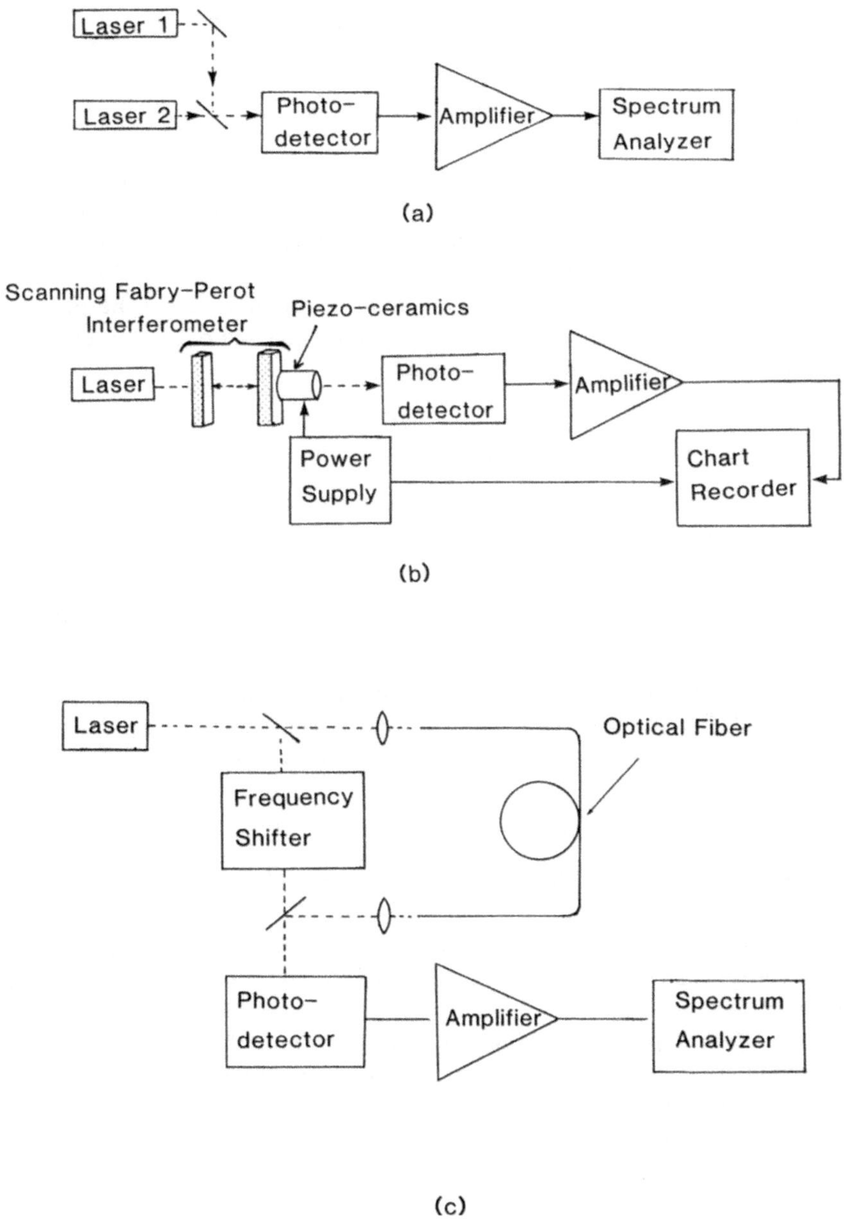

Fig. 4.1. Measurement methods of frequency fluctuation [C]. (a) Measuring the beat signal between two independent lasers. (b) Using Fabry-Perot interferometer. (c) Using an optical fiber as a delay line. (a) and (b) are used for measuring the power spectral density of FM noises and the field spectral profiles, respectively. (c) is for the field spectral profile.

laser is required, being different from the case in (a), one should note that the limits of resolution, sensitivity and measurement bandwidth are imposed by the performances of the measurement setups.

4.2 Magnitude of Laser Quantum Noise

It was described in 1.2 that a laser is a quantum device utilizing the phenomenon of stimulated emission and possessing fluctuations induced by a spontaneous emission even when light is in a minimum uncertainty state, i.e. in a coherent state. The magnitude of this quantum noise is estimated in the following [7]. In the case of photon emission from an atom, the initial and final states of the photon are represented by $|n\rangle$ and $|n+1\rangle$, respectively, thus, the rate of photon emission is expressed by using the creation operator a^+ as

$$\left|\langle n+1|a^+|n\rangle\right|^2 = n+1 \tag{4-20}$$

The rate of absorption can also be derived and is given by

$$\left|\langle n-1|a|n\rangle\right|^2 = n \tag{4-21}$$

By comparing these two expressions and by noting that the rates of stimulated emission and absorption are equal, one can easily find that the second term on the right-hand side of Eq.(4-20) corresponds to the rate of spontaneous emission, i.e.

$$\begin{aligned}&[\text{rate of spontaneous emission}]\\&= [\text{rate of stimulated emission induced by one photon}]\end{aligned} \tag{4-22}$$

Based on this relation, obtained by the quantum property of light, the magnitude of quantum noise induced by spontaneous emission is estimated by the classical model of an amplifier.

As the first step, the changing rate of a photon number is estimated when a photon with an energy of $\hbar\omega$ propagates through an active medium. Since this is induced by stimulated emission, spontaneous emission and absorption, which take place at every position along the propagation direction (the z-axis, see Fig. 4-2), the rate equation for the photon number $n(z)$ is given by

$$\mathrm{d}[n(z)\hbar\omega]/\,\mathrm{d}z = \alpha_G n(z)\hbar\omega + P_s \tag{4-23}$$

where P_s is the noise energy induced by spontaneous emission. α_G is the gain per

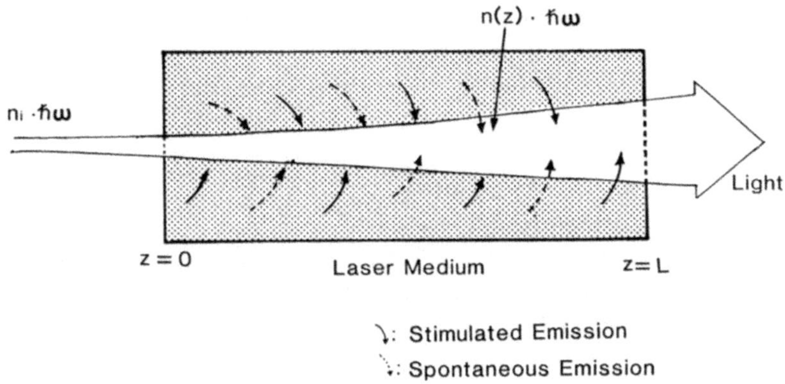

Fig. 4.2. Schematic explanation for the light amplification in a laser medium.

unit length due to stimulated emission, which is equal to $2n_q/c$ times the gain per unit time α_q of the electric field amplitude of light (see Eq.(2-85)), where n_q and c are the refractive index of the active medium and the speed of light in vacuum, respectively. Therefore, α_G is given by

$$\alpha_G = \frac{2n_q}{c}\frac{\omega|p|^2}{2\varepsilon_0\hbar\gamma}(N_a - N_b) \equiv \kappa(N_a - N_b), \qquad (4-24)$$

where κ is a proportional constant. N_a and N_b are the populations of the upper and lower atomic levels, respectively. Equation (4-22) is thus transformed to

$$\hbar\omega[\mathrm{d}n(z)/\mathrm{d}z] = \kappa N_a n(z)\hbar\omega - \kappa N_b n(z)\hbar\omega + P_s, \qquad (4-25)$$

where the first and second terms on the right-hand side of this equation represent the stimulated emission from the upper atomic level and absorption from the lower atomic level, respectively.

It can be easily found from Eq.(4-22) that the value of P_s corresponds to the value of the first term of Eq.(4-25) by fixing $n(z) = 1$, i.e., $P_s = \kappa N_a \hbar\omega$. By substituting this result into Eq.(4-25), one has

$$\hbar\omega\frac{\mathrm{d}n(z)}{\mathrm{d}z} = \kappa(N_a - N_b)n(z)\hbar\omega + \kappa N_a\hbar\omega. \qquad (4-26)$$

The solution of this differential equation with the boundary condition $n(0) = n_i$ gives an output photon energy at $z = l$, which is given by

$$\hbar\omega n(l) = \hbar\omega n_{\mathrm{i}} G + \hbar\omega \frac{N_a}{N_a - N_b}(G-1) \tag{4-27}$$

where

$$G \equiv \exp\left\{\kappa\left(N_a - N_b\right)l\right\} \tag{4-28}$$

The second term of Eq.(4-27) implies that noise power appears at the output of the amplifier as a result of the amplification of spontaneous emission being independent of the input energy $\hbar\,\omega n_i$. The equivalent input noise energy is estimated by dividing this term by the gain G, i.e.

$$W_s = \hbar\omega \frac{N_a}{N_a - N_b}\frac{G-1}{G} \cong \hbar\omega \frac{G}{N_a - N_b} \tag{4-29}$$

It corresponds to the spontaneous emission energy. However, since the steady-state value of population inversion is clipped to its threshold value, as is pointed out by Eq.(A-13) in the Appendix, the denominator in this equation is replaced by the threshold value $(N_a\text{-}N_b)_{\mathrm{th}}$ in the following discussion.

The magnitude of quantum noise in the laser can be estimated by adding the electric field amplitude $\Gamma(t)$ of this spontaneously emitted light to the Van der Pol's equation (Eq.(2-83)), i.e.

$$\frac{dE_q}{dt} - E_q\left(-\Gamma_q + \alpha_q - \beta_q E_q^2\right) = \Gamma(t) \tag{4-30}$$

This equation, obtained by adding an additive noise term to Van der Pol's equation, is called a Langevin's equation and the added noise term is called Langevin's force.

It can be assumed that the temporal fluctuations rate of the spontaneous emission power is faster than the rising time of laser oscillation (approximately $1/(\alpha_q\text{-}\Gamma_q)$, see Eq.(2-83)) and that spontaneous emission takes place at a random time. It can be concluded from these assumptions that the statistical properties of $\Gamma(t)$ are represented by white noise, i.e. its two-sided power spectral density $S_{\Gamma t}(\omega')$ is constant ($\omega' = 2\pi f$, where f is the Fourier frequency) and the time average $\langle\Gamma(t)\rangle$ is zero. By noting these statistical properties of $\Gamma(t)$, the value of $S_{\Gamma t}(\omega')$ can be obtained. That is, the value of $S_{\Gamma t}(\omega')$ is obtained by noting that the fluctuating energy of the electric field amplitude is given by Eq.(4-29): The behavior of this fluctuating amplitude $a_n(t)$ is expressed in Eq.(4-30) as

$$\mathrm{d}a_n / \mathrm{d}t + \Gamma_q a_n = \Gamma(t) \tag{4-31}$$

where the gain due to the stimulated emission $\alpha_q - \beta_q E_q^2$ has been removed from Eq.(4-30). The Fourier transform of this relation could be given by the two-sided power spectral density $S_{at}(\omega')$ as a function of $S_{\Gamma t}(\omega')$, which is expressed as

$$S_{at}(\omega') = \frac{1}{\omega'^2 + \Gamma_q^2} \cdot S_{\Gamma t}(\omega'). \qquad (4-32)$$

The integration of this function over $-\infty < \omega' < \infty$ is proportional to the noise energy, i.e.

$$\overline{a}_n^2 = \int_{-\infty}^{\infty} S_{at}(\omega') d\omega' = \frac{\pi}{\Gamma_q} \cdot S_{\Gamma t}(\omega'). \qquad (4-33)$$

Noise energy is given by $\varepsilon_0 a_n^2/2$ and corresponds to the value of Eq.(4-29). Thus, one has

$$\hbar\omega \frac{N_a}{(N_a - N_b)_{\text{th}}} = \frac{\varepsilon_0}{2} \frac{\pi}{\Gamma_q} S_{\Gamma t}(\omega'), \qquad (4-34)$$

and therefore,

$$S_{\Gamma t}(\omega') = \frac{2\Gamma_q}{\pi\varepsilon_0} \hbar\omega \frac{N_a}{(N_a - N_b)_{\text{th}}}. \qquad (4-35)$$

By noting that the power spectral densities of the real and imaginary parts of $\Gamma(t)$, $\Gamma_r(t)$ and $\Gamma_i(t)$, are the half of Eq.(4-35) because $\Gamma(t)$ represents a randomly fluctuating noise term, they are given by

$$S_{\Gamma_r t}(\omega') = S_{\Gamma_i t}(\omega') = \frac{\Gamma_q}{\pi\varepsilon_0} \hbar\omega \frac{N_a}{(N_a - N_b)_{\text{th}}}. \qquad (4-36)$$

The magnitude of amplitude fluctuation (AM noise) can be derived from the results obtained above and by using Langevin's equation. For this purpose, the fluctuating amplitude and phase (a_n and ϕ_n) are introduced into the electric field amplitude, i.e.

$$E_q(t) = (A_0 + a_n)\exp(i\phi_n), \qquad (4-37)$$

where A_0 is the average value of the amplitude. The time averages of these fluctuations are

$$\langle a_n \rangle = \langle \phi_n \rangle = 0, \tag{4-38}$$

because they are randomly fluctuating quantities. By noting that $A_0 >> a_n$ after substituting Eq.(4-37) into Eq.(4-30), one has

$$\frac{\mathrm{d}a_n}{\mathrm{d}t} + iA_0 \frac{\mathrm{d}\phi_n}{\mathrm{d}t} + A_0\left[\Gamma_q - \alpha_q + \beta_q\left(A_0^2 + 2A_0 a_n\right)\right] = \Gamma_q e^{-i\phi_n}. \tag{4-39}$$

By taking the time average of this relationship, one can easily find that the quantity in the bracket must be zero because $\langle a_n \rangle = \langle \phi \rangle = \langle \Gamma(t) \rangle = 0$. This requirement gives

$$A_0^2 = \left(\alpha_q - \Gamma_q\right) / \beta_q, \tag{4-40}$$

which is exactly equal to Eq.(2-90). After substituting this expression into Eq.(4-39) and separating real and imaginary parts, one has

$$\frac{\mathrm{d}a_n}{\mathrm{d}t} + 2A_0^2 \beta_q a_n = \Gamma_r(t), \tag{4-41}$$

and

$$A_0 \frac{\mathrm{d}\phi_n}{\mathrm{d}t} = \Gamma_i(t). \tag{4-42}$$

The magnitude of AM noise is obtained from Eq.(4-41), i.e. the Fourier transform of Eq.(4-41) gives the two-sided power spectral density of the AM noise

$$S_{at}(\omega') = \frac{1}{\omega'^2 + \left(2A_0^2\beta_q\right)^2} \frac{\Gamma_q}{\pi\varepsilon_0} \hbar\omega \frac{N_a}{\left(N_a - N_b\right)_{\mathrm{th}}}. \tag{4-43}$$

The profile of this power spectral density is a Lorentzian, as shown in Fig. 4-3, and the d.c. value at $\omega' = 0$ is

$$S_{at}(0) = \frac{1}{\left(\overline{N} / \overline{N}_{\mathrm{th}} - 1\right)^2 \Gamma_q} \frac{\hbar\omega}{4\pi\varepsilon_0} \frac{1}{\left(N_a - N_b\right)_{\mathrm{th}}} \propto \frac{1}{\left(\overline{N} / \overline{N}_{\mathrm{th}} - 1\right)^2}, \tag{4-44}$$

in which Eqs.(2-91) and (4-40) are used. This expression signifies that the D.C. value is inversely proportional to the square of the normalized bias level $\left(\overline{N}/\overline{N}_{\text{th}}-1\right)$. The HWHM (half width at half maximum) ω_h' is given by

$$\omega'_h = 2A_0^2\beta_q = 2\Gamma_q\left(\overline{N}/\overline{N}_{\text{th}}-1\right), \qquad (4-45)$$

which is proportional to the normalized bias level. The total energy of the AM noise is

$$\varepsilon_0\overline{a}_n^2/2 = \frac{\varepsilon_0}{2}\int_{-\infty}^{\infty} S_{at}(\omega')\mathrm{d}\omega' = \frac{\hbar\omega}{4}\frac{N_a}{\left(N_a-N_b\right)_{\text{th}}}\frac{1}{\overline{N}/\overline{N}_{\text{th}}-1}, \qquad (4-46)$$

which is inversely proportional to the normalized bias level. Therefore, it can be concluded from these relations that the laser operating at the higher bias level shows a lower AM noise energy even though the noise bandwidth is wider.

The two-sided power spectral density of the angular frequency fluctuation $\mathrm{d}\phi_n/\mathrm{d}t$ is derived from Eq.(4-42), which is given by

$$S_{vt}(\omega') = \frac{1}{A_0^2}\cdot S_{\Gamma_i t}(\omega') = \frac{\Gamma_q}{\pi\varepsilon_0 A_0^2}\hbar\omega\frac{N_a}{\left(N_a-N_b\right)_{\text{th}}}, \qquad (4-47)$$

where $\varepsilon_0 A_0^2/2$ represents the energy of the laser field stored in the cavity. The

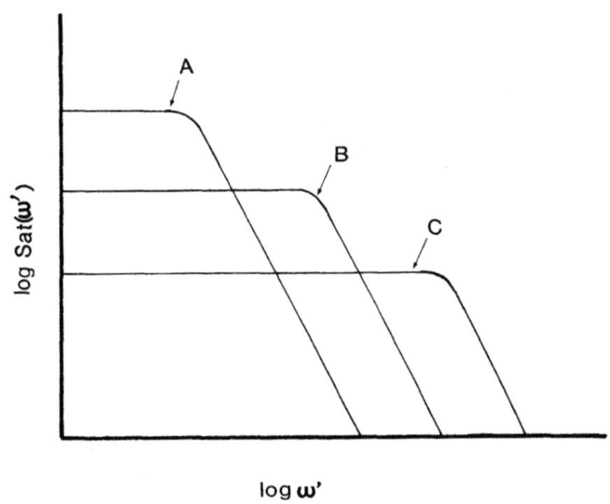

Fig. 4.3. Power spectral density of the AM noise of a laser (see (4.43)). Normalized bias level is increased in the order of curves A, B and C.

intracavity power of the laser P is derived by multiplying the dissipation rate of this energy from the active medium to the external reservoir, i.e.

$$P = 2\Gamma_q\left(\varepsilon_0 A_0^2 / 2\right) = \Gamma_q \varepsilon_0 A_0^2. \qquad (4-48)$$

By using this expression, Eq.(4-47) is transformed into

$$S_{vt}(\omega') = \frac{\Gamma_q^2}{\pi P}\hbar\omega \frac{N_a}{\left(N_a - N_b\right)_{\text{th}}}. \qquad (4-49a)$$

For comparison with experimental results, the one-sided power spectral density per unit width of the Fourier frequency range is obtained from this expression as

$$S_v(f) = \frac{4\Gamma_q^2}{P}\hbar\omega \frac{N_a}{\left(N_a - N_b\right)_{\text{th}}}. \qquad (4-49b)$$

The one-sided power spectral density of the normalized angular frequency fluctuation ($y \equiv (\mathrm{d}\phi_n/\mathrm{d}t)/\nu_{ql}$) is obtained from this expression as

$$S_y(f) = \frac{4\Gamma_q^2}{\nu_{ql}^2 P}\hbar\omega \frac{N_a}{\left(N_a - N_b\right)_{\text{th}}}. \qquad (4-49c)$$

These three expressions represent the characteristics of white noise determined by these of spontaneous emission. That is, it can be regarded as white noise below the Fourier frequency corresponding to the relaxation rate of the dipole moment induced in the atom (= several 10 THz).

It can easily be found by comparing Eqs.(4-16) and (4-49) that the field spectral profile is a Lorentzian whose FWHM is

$$\Delta f = \frac{\pi\left(\Delta\nu'_c\right)^2}{P}\hbar\omega \frac{N_a}{\left(N_a - N_b\right)_{\text{th}}}, \qquad (4-50)$$

where the cavity loss Γ_q was replaced by the FWHM of the cavity resonance spectral shape $\Delta\nu_c$ and further replaced by $\Delta\nu_c' = \Delta\nu_c/2\pi$ to represent the unit of Hertz. Equation (4-50) shows that the FM noise magnitude is proportional to the square of the cavity loss. Furthermore it is also inversely proportional to the intra-cavity laser power, which corresponds to the relationship shown in Eq.(1-71). Here, it is easily shown that the intra-cavity laser power P is related to the laser output power

P_0 extracted from one of the cavity mirrors and is expressed as

$$P_0 = \frac{P}{2} \frac{\frac{1}{L} \ln\left(\frac{1}{R}\right)}{\alpha_{\mathrm{tl}}} \qquad (4-51)$$

where the reflectivities of the two cavity mirrors are assumed to be R. Thus, Eq.(4-50) can be transformed by using Eqs.(2-10) and (4-51) to

$$\Delta f = \frac{\hbar\omega}{8\pi P_0}\left(\frac{c}{n_q L}\right)^2 \left(\alpha_l L + \ln\frac{1}{R}\right)\left(\ln\frac{1}{R}\right) \cdot \frac{N_a}{\left(N_a - N_b\right)_{\mathrm{th}}} \qquad (4-52)$$

The expression given by this equation is called a Schawlow-Townes formula [8].

The FWHM of a He-Ne laser with several mW output power is as narrow as about several mHz; however, that of a semiconductor laser is about several 100 kHz because of its large cavity loss. It corresponds to the magnitude of FM noise of the coherent state of light determined by the magnitude of spontaneous emission fluctuation. However, the FM noise magnitudes of actual laser devices are usually far larger than this quantum noise because of several additional external noise sources. These additional noises are described in the next section.

4.3 Additional Noises in Lasers

4.3.1 Gas, dye and solid state lasers

Equation (2-97) shows that fluctuations in the refractive index n_q and the cavity length L can induce additional FM noise since the angular frequency of the laser oscillation ν_{ql} is close to the cavity resonance angular frequency ν_q. For example, the additional FM noises of an He-Ne laser are induced by external noise sources caused by air flows in the laser cavity, mechanical vibrations, plasma instability in discharged He-Ne gas, noises in current source and ambient temperature fluctuations. The magnitudes of these FM noises are usually much larger than those of the quantum noise given in 4.2. Examples of these FM noise magnitudes are represented by the Allan variance and are shown by Fig. 4-4 [9]. Curve A represents quantum noise. Curves B and C are the additional noises induced by plasma instability and fluctuations in refractive index, respectively. The total magnitude of FM noise is given by their sum and is represented by curve D. The magnitude of FM noise in an Ar^+ laser is larger than that in an He-Ne laser because of the large fluctuations in refractive index and temperature induced by the large discharge current, temperature fluctuations by water-cooling and decreases in gas pressure by sputtering [10]. The magnitude of plasma instability in a CO_2 laser is usually smaller than those in the two atomic lasers shown above because vibration-rotation

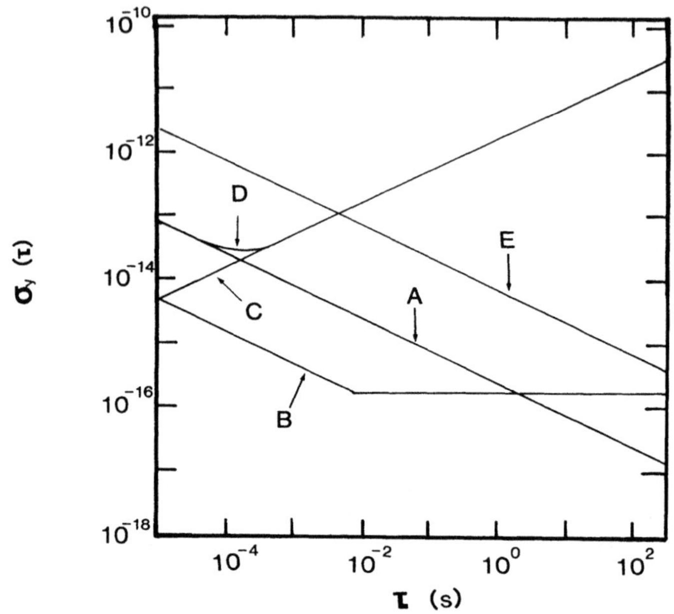

Fig. 4.4. Calculated results of the square root of the Allan variance of the frequency fluctuation of a He-Ne laser (after Ohtsu, Fukada, Tako and Tsuchida [9]). A: Quantum noise resulted from spontaneous emission. B: Fluctuation of current source. C: Fluctuation of ambient temperature. D: Fluctuation under the free-running condition (summatiom of the values in A, B and C). E: Shot-noise limit of the frequency-controlled laser.

transitions in molecular gases are used. The principal FM noise sources in CO_2 lasers are the frequency drift induced by cavity length drift due to temperature fluctuations.

In dye lasers, the principal noise sources are fluctuations in the optical pumping power, in the thickness and flow rate of the dye jet and in the micro-bubbles contained in the dye jet. To reduce the size and number of these bubbles, several techniques have been developed to design an optimal nozzle structure for the dye jet and a filter to purify the dye solution, and to add some solvents, e.g., ethylene glycol, to increase viscosity.

The principal FM noise sources in a conventional YAG laser are the power fluctuations in an arc lamp and temperature fluctuations. However, recently developed semiconductor laser-pumped YAG lasers have a much smaller magnitude of temperature fluctuations induced by the pumping source. A very stable laser oscillation has been realized by NPRO (Fig. 3-14(a)) because the YAG crystal itself forms a laser cavity to reduce the contributions from additional external noise sources.

4.3.2 Semiconductor lasers [11]

Semiconductor lasers exhibit specific noise characteristics because their cavity loss is large, the adiabatic approximation of carrier density variation is not valid due to the large and fast carrier density variation, and the width of the gain spectral profile is more than 100 times larger than the longitudinal mode separation. Especially, because of the very wide gain in spectral profile, there are always unwanted vestigial side modes even in a longitudinal mode controlled semiconductor laser; consequently, the concept of single longitudinal mode semiconductor lasers can only be approximated. If the present discussion is limited only to two longitudinal modes in order to simplify the analyses, Eqs.(2-100) and (2-101) can be used. Since the coefficients' values in these equations depend on the carrier density, the equation of motion for the carrier density $n^{(0)}$ is required for these analyses and is expressed as

$$\mathrm{d}n^{(0)} / \mathrm{d}t = n_q \sqrt{\frac{\varepsilon_0}{\mu_0}} \left(\frac{2}{\hbar \nu_{1l}} \alpha_1 E_1^2 + \frac{2}{\hbar \nu_{2l}} \alpha_2 E_2^2 \right) - \frac{n^{(0)}}{\tau_s} + \frac{I}{Ve} \qquad (4-53)$$

The gain saturation is caused by the intraband relaxation of carriers in the conduction band, whose relaxation constant τ_{in} is as short as 0.3 ps. The gain saturation coefficients (β_q and θ_{qp} in Eq.(2-100)) are proportional to $(n^{(0)})^2$. In addition, when the carrier density $n^{(0)}$ decreases due to the transition from conduction to valence bands, as is shown in Eq.(4-53), gain saturation also occurs because the linear gain α_q depends on $n^{(0)}$. The time constant of this relaxation is given by τ_s ($\cong$ 3 ns) which is much longer than τ_{in}. However, the characteristics of the mode hopping and mode partition phenomena described below depend mainly on intraband relaxation. That is, these phenomena are quantum phenomena driven by spontaneous emission fluctuations coupled to each longitudinal mode and based on gain saturation due to the intraband relaxation of carriers.

The characteristics of an index-guided laser, not the longitudinal mode controlled laser, are discussed for simplicity of analysis. The linear gains of the two longitudinal modes can be set almost equal (i.e. $\alpha_1 \cong \alpha_2$) by adjusting the injection current and temperature. Figure 4-5(a) shows the time dependences of each mode intensity observed under this condition, which shows a clear switching phenomenon [12] called mode hopping. This is attributed to the suppression of one mode due to the decrease in carrier density because it is spent in lasing the other mode. Since this phenomenon is driven by randomly generated spontaneous emissions which work as triggering forces in lasing the other mode, the intensity fluctuations (IM noise) follow the statistics of a Poisson process. Therefore, the profile of the power spectral density of the IM noise can be approximated as a Lorentzian whose cutoff frequency represents the average frequency of occurrence of mode hopping. It has been confirmed experimentally that the cutoff frequency decreases exponentially with the increasing normalized bias level I/I_{th} (the injection current I divided by its threshold value I_{th}) [13].

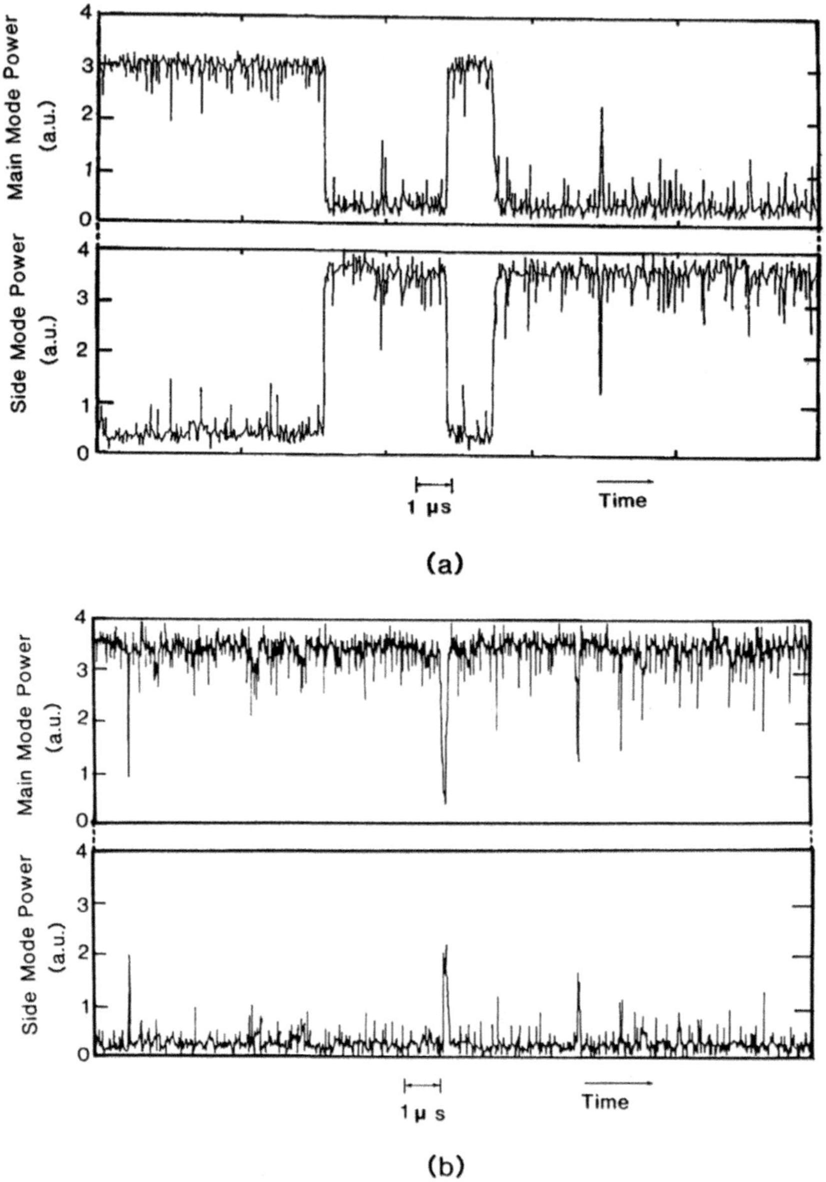

Fig. 4.5. Measured results of temporal variations of the mode powers of the semiconductor laser when two longitudinal modes oscillate (after Ohtsu and Teramachi [12], ©1989 IEEE). (a) Phenomenon of mode hopping. (b) Phenomenon of mode partition.

On the other hand, a nearly-single longitudinal mode oscillation can be realized, i.e. the linear gain of one mode can be fixed far larger than that of the others ($\alpha_1 >> \alpha_2 > \Gamma_1, \Gamma_2$) by fixing the injection current and temperature at other values. Figure 4-5(b) shows time dependences of two mode powers [12]. The transient decrease in the main mode power is called a power drop-out. The power exchange between the main- and sub-modes, shown in this figure, is called mode partition. These IM noises occur from the same origin as those in mode hopping, i.e. the gain of the main mode decreases with a decrease in carrier density due to the transient increase in sub-mode power driven by spontaneous emission. It has been confirmed experimentally that the pulse width of the power drop-out, the probability of occurrence of power drop-out and the variance of power fluctuations decrease exponentially with increases in I/I_{th}.

The characteristics of mode hopping and mode partition can be commonly analyzed by Langevin's two-mode equation by adding Langevin's forces to represent mutually uncorrelated spontaneous emissions in Eq.(2-100). However, the treatment of these equations is not straightforward because they are two-dimensional and nonlinear differential equations. To overcome these difficulties, they are transformed to linear Fokker-Planck equations describing the probability density $p(I_1, I_2, t)$, where I_1 and I_2 are the intensities of the two modes at time t [12, 13]. In this analysis, it is assumed that the magnitude of temporal fluctuations of the carrier density described in Eq.(4-53) can be neglected because the sum of the two mode powers is maintained constant, which can be realized experimentally by fine adjustments of the injection current and temperature even when mode hopping and partition occur. A Fokker-Planck's equation can be derived from Langevin's equations and is a linear eigenvalue equation whose minimum eigenvalue is zero. The eigenfunction corresponding to this eigenvalue gives the steady state solution

$$p_s(I_1\ I_2) = \frac{1}{B}\exp\left[-U(I_1\ I_2)\right] \qquad (4-54)$$

where B is a normalization constant. $U(I_1, I_2)$ is called a potential by analogy with classical thermo-dynamics, which depends on linear gains and gain saturation coefficients. Figures 4-6(a) and (b) show the contours of potentials in mode hopping and mode partition, respectively, drawn by substituting the numerical values of several experimental parameters corresponding to Figs. 4-5(a) and (b) [12]. The contour of (a) has its two minima M_1 and M_2 separated by a saddle point S. Both of these minima correspond to a highly probable state in which one of the two mode intensities is zero while the other is nonzero. The steady-state is described by one of these minima. Fluctuations by spontaneous emission can drive the representation point in phase space from one minimum to the other at random times. The intensity of each mode therefore tends to jump randomly between zero and nonzero values, which corresponds to the mode hopping phenomena shown in Fig. 4-5(a). The reciprocal of the second eigenvalue in Fokker-Planck's equation represents the

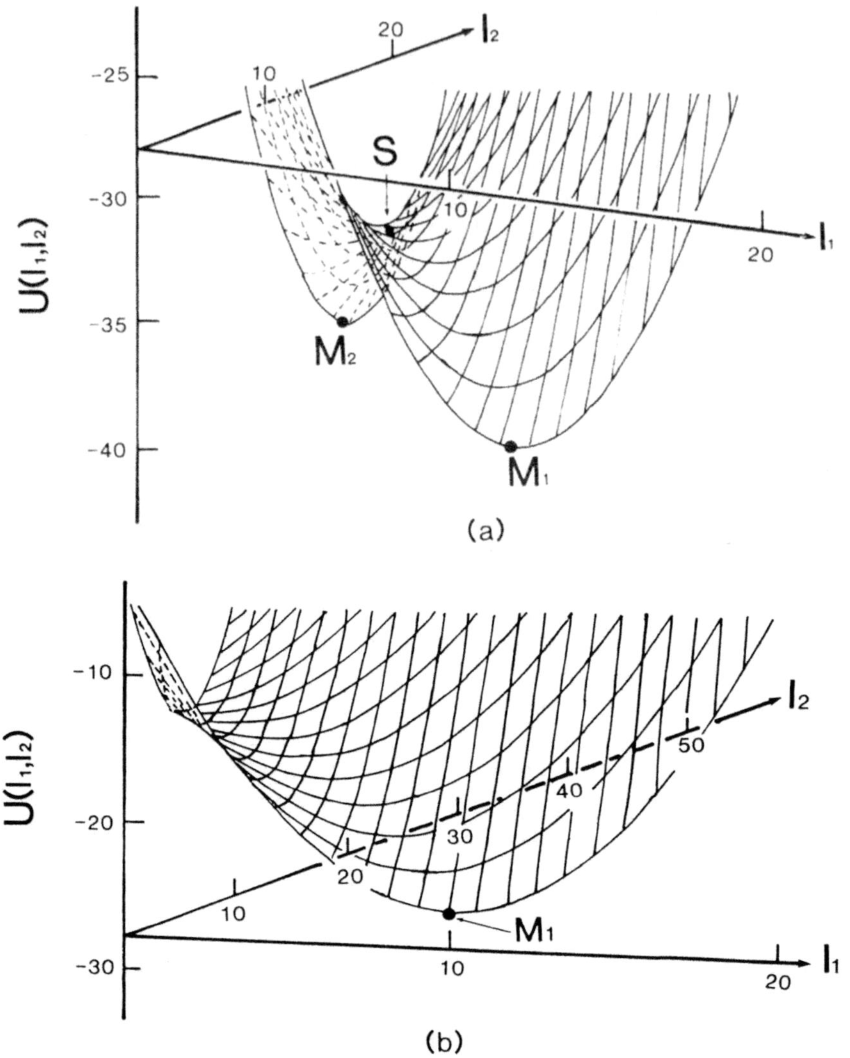

Fig. 4.6. Calculated result of the contours of the potential $U(I_1, I_2)$ (after Ohtsu and Teramachi [12], ©1989 IEEE). (a) Corresponding to the mode hopping. (b) Corresponding to the mode partition. Points M_1 and M_2 represent the positions of the potential minima.

frequency of mode hopping. The contour in (b) has one minimum (M_1), which corresponds to the single longitudinal mode oscillation in the steady-state. However, fluctuations by spontaneous emission can deviate transiently the representation point from M_1; and subsequently, back to M_1. This dynamic behavior corresponds to the phenomena of power drop-out and mode partition. Experimental results have been quantitatively described by this theoretical model. Furthermore, it has been pointed out that the stochastics of the IM noise around the threshold were analogous to the critical slowing down around the critical point in the phase transition phenomenon [12].

The phenomena described above are attributed to a strong coupling between the two modes, which is expressed by the fact that the coupling constant, defined by the ratio between the self- and cross- saturation coefficients as

$$C \equiv \theta_{12}\theta_{21} / \beta_1\beta_2 \qquad (4-55)$$

and C is larger than unity. This is due to a very short intraband relaxation time τ_{in}. In contrast to semiconductor lasers, most gas lasers show weak coupling, i.e. $C <$ 1, which means that each mode lases independently.

Since the frequency of mode hopping decreases with increases in the magnitude of cross saturation, the use of a saturable absorber has been tried by doping an impurity (e.g., Te) into the clad layer to form a trap-level for the electron, i.e. the DX center [14, 15]. Phenomenon of power drop-out can be observed even in longitudinal mode controlled semiconductor lasers. That is, even the DFB and DBR lasers show that they are not single longitudinal mode lasers in a strict sense. It has been estimated that the ratio between steady-state intensities of the main and side modes in these lasers should be larger than 40 dB to keep the bit error rate due to a power drop-out lower than 1×10^{-9} when these lasers are used in optical communication systems [16, 17]. It has been demonstrated that the side mode power can be decreased by suppressing spatial hole burning by segmenting the electrodes [18].

Figure 4-7 shows a schematic configuration in which another serious IM noise occurs, i.e., it has been empirically known that laser oscillation became unstable when reflected light was injected into the laser after having reflected back from an

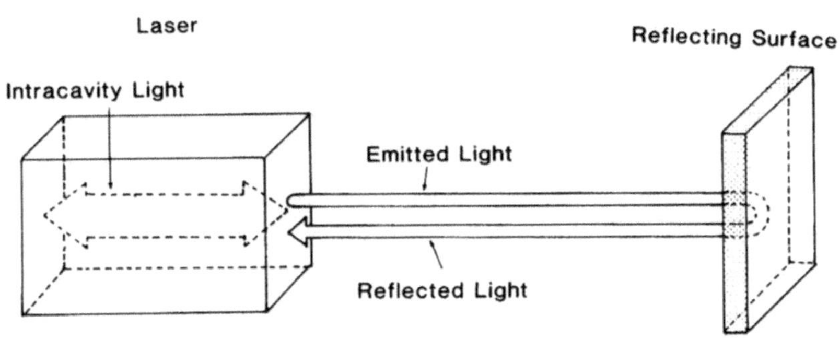

Fig. 4.7. Schematic explanation for noise generation mechanism induced by injection of the reflected light [C].

external reflecting surface [C]. The probability of occurrence of this instability, induced by an injection of phase-delayed light, is large in semiconductor lasers because their facet reflectivities are low and carriers are easily optically pumped by the injected light. Figure 4-8 shows results of computer simulations to estimate increases in IM noise by increases in reflected light intensity [19], which was derived by adding a term representing the electric field amplitude of the reflected light to Van der Pol's equation. It is easily seen from this figure that the effect of reflected light appears even when 0.003% of emitted intensity is reflected back into the laser, and finally, at 5%, a pulsive oscillation takes place.

This phenomenon is the deterministic instability which occurs when a phase-delayed lightwave is injected into the self-sustained oscillator, i.e. a CW laser, and finally, leads to chaos [20]. In an actual system, randomly fluctuating temperature,

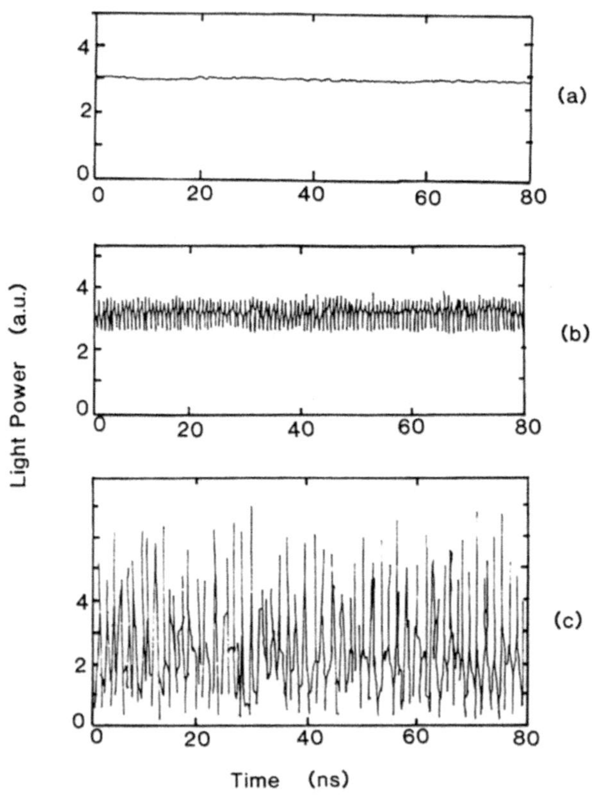

Fig. 4.8. Calculated result of the temporal power variation of a semiconductor laser caused by the injection of a reflected light (after Ohtsu [19]). The distance between output end and reflection surface is 5 cm. The ratio between the emitted power and the injected power after reflection is 0% (a), 0.003% (b), and 5% (c), respectively.

injection current and the position of the external reflector could make the characteristics of this instability very complicated and induce further increases in magnitude of instability. It is necessary to increase the laser facet reflectivity or to use an optical isolator. An optical isolator is equivalent to the Faraday rotator in Fig. 3-9, which utilizes the non-reciprocal rotation of the polarization of light induced by the Faraday effect (i.e. one of the magneto-optical effects). It has been empirically shown that the optical isolation of about 60 dB is required to reduce the effect of reflected lightwave in practical application systems.

The following discussions concentrate on the characteristics of noises in a single longitudinal mode laser (more strictly, a nearly single longitudinal mode laser) by assuming that the additional noises presented above have been removed. In addition to the quantum noise in Fig. 4-3, IM noise could be induced by carrier density fluctuations due to optical pumping by spontaneously emitted light. Therefore, this carrier-induced IM noise can be regarded also as quantum noise. Figure 4-9 shows the power spectral density of these noises, in which the effect of carrier density fluctuations gives a resonant peak at the relaxation oscillation frequency [11]. And this relaxation oscillation frequency corresponds to the high frequency cutoff. Furthermore, these carrier density fluctuations could induce current fluctuations which could also induce temperature fluctuations due to the self-heating of the laser device. This can also be regarded as quantum noise. In

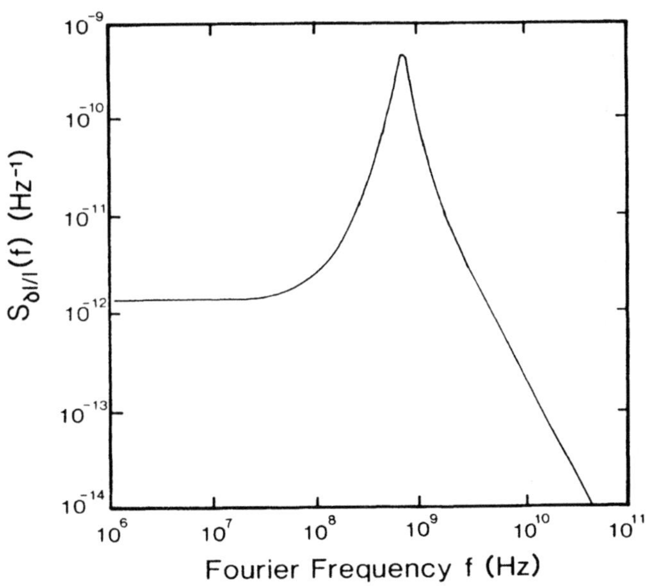

Fig. 4.9. Power spectral density of the IM noise of a single longitudinal mode semiconductor laser (after Ohtsu and Tako [11]).

addition to these, there are also IM noises due to current fluctuations in the current source and fluctuations in the threshold current or quantum efficiency due to ambient temperature fluctuations.

Similar discussions can also be given for the characteristics of FM noise, i.e. quantum noise is composed of spontaneous emission, carrier density fluctuation and temperature fluctuations. That is, in addition to the quantum noise given in Eq.(4-49), the longitudinal mode frequency shown in Eq.(2-7) could fluctuate due to fluctuations in the refractive index caused by fluctuations in the carrier density, which are optically pumped by spontaneous emission. The ratio between the power spectral density due to carrier-induced fluctuations and that in Eq.(4-49) is represented by α^2, where α has been called the α-parameter or the linewidth enhancement factor [21]. IM and FM noises are correlated with each other via carrier density fluctuations. That is, the α-parameter is given by the ratio between changes in real and imaginary parts of the complex susceptibility (corresponding to Eq.(2-69)) $\Delta\chi'$ and $\Delta\chi''$

$$\alpha \equiv \Delta\chi'/\Delta\chi'', \qquad (4-56)$$

where $\Delta\chi'$ and $\Delta\chi''$ correspond to the change in refractive index and that in gain of active medium. It has been reported that the α-parameter took a value within a range

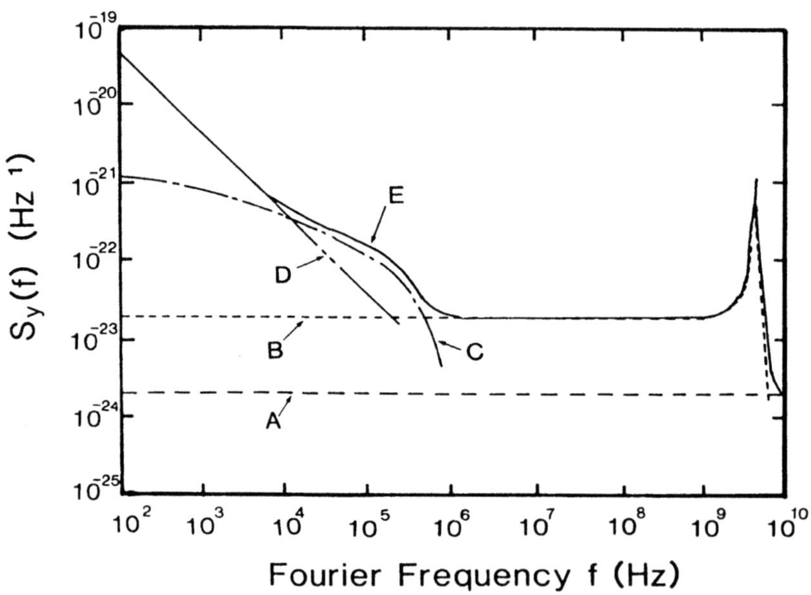

Fig. 4.10. Power spectral density of the quantum FM noise of a single longitudinal semiconductor laser. A: Fluctuation of spontaneous emission. B: Fluctuation of carrier density. C: Fluctuation of temperature. D: $1/f$ fluctuation. E: Total quantum FM noise.

of 2 ~ 9 [22]. Center frequencies of the spectral profiles of χ' and χ'' are different from each other because the gain spectrum has a complicated profile due to the intrinsic band structure of the semiconductor energy level. Therefore, the value of α depends on the material of the active medium, waveguide structure, and so on. Figure 4-10 shows the power spectral density of the FM noise. The contribution of the carrier density fluctuations shows a resonant peak at the relaxation oscillation frequency, which corresponds to the high frequency cutoff. Below this frequency, the ratio between the values due to carrier density fluctuations and spontaneous emission corresponds to α^2. In addition to these contributions, slowly fluctuating components are observed due to the drift in longitudinal mode frequency. This is because self-heating by current fluctuations is induced by carrier density fluctuations. These contributions should also be regarded as quantum noises. The field spectral profile due to these quantum noises shows FM sidebands due to the relaxation oscillation of the carrier. Furthermore, the half linewidth is broaden by the factor α^2 as compared to that of the spontaneous emission only. Thus, the half linewidth of the semiconductor laser is expressed as

$$\Delta f_{\mathrm{LD}} = \left(1+\alpha^2\right)\Delta f \qquad (4-57)$$

where Δf is the linewidth due to the spontaneous emission given in Eq.(4-52), in which the quantity $N_a/(N_a\text{-}N_b)_{\mathrm{th}}$ should be replaced by the spontaneous emission factor n_{sp} in semiconductor lasers. This factor is expressed as

$$n_{\mathrm{sp}} = 1/\left[1-\exp\left\{\left(\hbar\nu_l + E_{\mathrm{Fv}} - E_{\mathrm{Fc}}\right)/k_B T\right\}\right] \qquad (4-58)$$

where E_{Fv} and E_{Fc} are the quasi-Fermi levels of the valence and conduction band, respectively, k_B is Boltzmann's constant, T is the temperature and the value of n_{sp} falls between 1 and 2. Equation (4-57) has been called a modified Schawlow-Townes formula for semiconductor lasers [21].

As a source of additional noise, flicker noise can sometimes be observed and is independent from laser power. This has been attributed to several origins such as fluctuations in carrier mobility [23]. However, its precise origin has not yet been fully identified. Due to this flicker noise, a power-independent linewidth of about 2-8 MHz can be observed even within the limit of infinite laser power ($P \to \infty$) [24].

Furthermore, as an external source of noise, current fluctuations from the current source and ambient temperature fluctuations could induce FM noise and are expressed as

$$\delta\nu = -\nu_l\left[\left(\frac{A}{n_q}\right)\delta N_c(I) + \left(\alpha_T+\beta_T\right)\left\{\delta T_1(I)+\delta T_2\right\}\right] \qquad (4-59)$$

The constants in this equation are exactly the same as those in Eq.(3-4). $\delta N_c(I)$, $\delta T_1(I)$ and δT_2 are the carrier density fluctuations due to current fluctuations, temperature fluctuations due to current fluctuations and the ambient temperature fluctuations,

Table 4.2. Variations of wavelength λ and frequency ($\nu/2\pi$) of semiconductor lasers with the changes of D.C. current and ambient temperature [C].

Variations of Wavelength and Frequency	AlGaAsLaser(0.8μm)	InGaAsLaser(1.5μm)
$\frac{d\lambda}{dI}$	+0.006nm/mA	+0.008nm/mA
$\frac{1}{2\pi}\cdot\frac{d\nu}{dI}$	-2.8GHz/mA	-1.1GHz/mA
$\frac{d\lambda}{dT'}$	+0.06nm/K	+0.08nm/K
$\frac{1}{2\pi}\cdot\frac{d\nu}{dT'}$	-28GHz/K	-11GHz/K

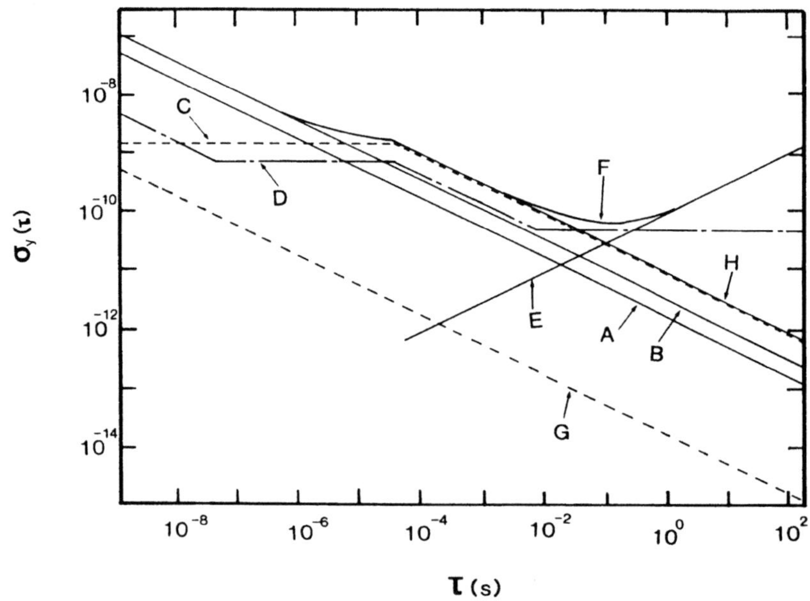

Fig. 4.11. Calculated results of the square root of the Allan variance of the FM noise of a single longitudinal mode semiconductor laser(after Ohtsu, Fukuda, Tako, and Tsuchida [9]). A: Fluctuation of spontaneous emission. B: Fluctuation of carrier density. C: Fluctuation of temperature. Curves A, B, and C correspond to the curves A, B,and C in Fig. 4.10, respectively. D: Fluctuation of current from a current source. E: Fluctuation of ambient temperature. F: FM noise under free-running condition. (summation of the values in from A to D). G: Shot-noise limit of the frequency-controlled laser. H: Total quantum fluctuation (summation of the values in A, B and C, and corresponds to the curve E in Fig. 4.10).

respectively. Table 4-2 summarizes the rates of frequency drift due to slow changes in current and ambient temperature [C]. Figure 4-11 shows estimated values of the square root of the Allan variance for the magnitude of all fluctuations described above [9]. In summary, the instantaneous frequency fluctuation at time t is given by

$$\delta\nu(t) = \frac{\mathrm{d}\phi_n}{\mathrm{d}t} = \Gamma_q\left(1+\alpha^2\right)\Gamma_s(t) + \Gamma_{ex}(t) \qquad (4-60)$$

which was derived from Eq.(4-42) to give a more concrete expression. The first term on the right-hand side of this equation gives quantum noise, in which the magnitude of the cavity loss Γ_q appears because Eq.(4-49) has shown that the magnitude of quantum noise is proportional to the cavity loss. $\Gamma_s(t)$ is the instantaneous value of spontaneous emission fluctuations. α is the α-parameter for the semiconductor laser, which may be neglected for gas lasers. The second-term, $\Gamma_{ex}(t)$, is the magnitude of external noises. The noises presented here give the total FM noise in semiconductor lasers.

[Textbooks]

A) P. Kartaschoff, *Frequency and Time*, Academic Press, London, 1978.

B) M. Ohtsu, *Lasers and Atomic Clocks*, Ohm-sha Publishing, Tokyo, 1986 (in Japanese).

C) M. Ohtsu, Technology for Light Emitting and Detecting Devices for Fiber Sensor, Chapter 3 in *Optical Fiber Sensor*, ed. by T. Okoshi, Ohm-sha Publishing, Tokyo, 1986 (in Japanese).

D) K. Yoshimura, Y. Koga, and N. Oura, *Frequency and Time*, Inst. Electronics, Communication and Electric Engineer of Japan, Tokyo, 1989 (in Japanese).

[References]

1) D. W. Allan, "Statistics of atomic frequency standards", *Proc. IEEE*, Vol. 54 (1966) pp. 221–230.

2) R. E. Beehler, R. C. Mockler, and J. M. Richardson, "Cesium beam atomic time and frequency standards", *Metrologia*, Vol. 1 (1965) pp. 114–131.

3) R. E. Beehler and D. J. Glaze, "The performance and capability of cesium beam frequency standards at the National Bureau of Standards", *IEEE Trans. Instrum. Meas.*, Vol. IM-15 (1966) pp. 48–55.

4) H. W. Rowe, *Signals and Noise in Communication Systems*, D. Van Nostrand, Princeton, 1965.

5) C. -H. Shin and M. Ohtsu, "Heterodyne Optical Phase-Locked Loop by Confocal Fabry-Perot Cavity Coupled AlGaAs Lasers", *IEEE Photonics Technol. Lett.*, Vol. 2 (1990) pp. 297–300.

6) T. Okoshi, K. Kikuchi and A. Nakayama, "Novel method for high resolution measurement of laser output spectrum", *Electron. Lett.*, Vol. 16 (1980) pp. 630–631.

7) K. Shimoda, T. Yajima, Y. Ueda, T. Shimizu and T. Kasuya, *Quantum Electronics*, Shokabo Publishing, Tokyo, 1972 (in Japanese).

8) A. Blaquierre, "Spectre d'un oscillateur maser; relation avec la theorie des autooscillateurs nonlineaires classiques", *Compt. Rend.*, Vol. 26 (1962) pp. 2929–2931.

9) M. Ohtsu, H. Fukada, T. Tako, and H. Tsuchida, "Estimation of the Ultimate Frequency Stability of Semiconductor Lasers", *Jpn. J. Appl. Phys.*, Vol. 22 (1983) pp. 1157–1166.

10) F. Spieweck, "Frequency Stabilization of Ar^+ Lasers at 582 THz Using Expanded Beams in External $^{127}I_2$ Cells", *IEEE Trans. Instr. and Meas.*, Vol. IM-29 (1980) pp. 361–363.

11) M. Ohtsu and T. Tako, "Coherence in Semiconductor Lasers", in *Progress in Optics XXV*, ed. E. Wolf, Elsevier, Amsterdam (1988) pp. 191–278.

12) M. Ohtsu and Y. Teramachi, "Analyses of Mode Partition and Mode Hopping in Semiconductor Lasers", *IEEE J. Quantum Electron.*, Vol. 25 (1989) pp. 31–38.
13) M. Ohtsu, Y. Teramachi, Y. Otsuka and A. Osaki, "Analysis of Mode-Hopping Phenomena in an AlGaAs Laser", *IEEE J. Quantum Electron.*, Vol.QE-22 (1986) pp. 535–543.
14) N. Chinone, T. Kuroda, T. Ohtoshi, T. Takahashi and T. Kajimura, "Mode hopping noise in index-guided semiconductor lasers and its reduction by saturable absorbers", *IEEE J. Quantum Electron.*, Vol. QE-21 (1985) pp. 1264–1269.
15) M. Ohtsu, Y. Teramachi, and T. Miyazaki, "Analyses of suppression of mode hopping in an AlGaAs laser by saturable absorber", *Opt. Commun.*, Vol. 61 (1987) pp. 203–207.
16) R. A. Linke, B. L. Kasper, C. A. Burrus, Jr., I. P. Kaminov, J. S. Ko, and T. P. Lee, "Mode power partition events in nearly single-frequency lasers", *J. Lightwave Technol.*, Vol. LT-3 (1985) pp. 706–712.
17) K. -Y. Liou, M. Ohtsu, C. A. Burrus, Jr., U. Koren, and T. L. Koch, "Power Partition Fluctuations in Two-Mode-Degenerate Distributed-Feedback Lasers", *J. Lightwave Technol.*, Vol. 7 (1989) pp. 632–639.
18) M. Usami, S. Akiba, and Y. Matsushima, "Mode characteristics of $\lambda/4$-shifted DFB lasers with the distributed current injection along the cavity", Extended Abstracts of the 49th Autumn Meeting of the Jpn. Soc. Appl. Phys., 1988 (Jpn. Soc. Appl. Phys., Tokyo, 1988) 6p-ZC-16 (in Japanese).
19) M. Ohtsu, "Noises in Lasers", *Physics Monthly*, Vol. 6 (1985) pp. 297–303 (in Japanese).
20) P. Berge, Y. Pomeau and C. Vidal, "*Orders within Chaos*", John Wiley & Sons, New York, 1984.
21) C. H. Henry, "Theory of the linewidth of semiconductor lasers", *IEEE J. Quantum Electron.*, Vol. QE-18 (1982) pp. 259–264.
22) M. Oshinski and J. Buus, "Linewidth broadening factor in semiconductor lasers — An overview", *IEEE J. Quantum Electron.*, Vol. QE-23 (1987) pp. 9–29.
23) M. Ohtsu and S. Kotajima, "Derivation of the Spectral Width of a 0.8 μm AlGaAs Laser Considering 1/f Noise", *Jpn. J. Appl. Phys.*, Vol. 23 (1984) pp. 760–764.
24) D. Welford and A. Mooradian, "Observation of linewidth broadeing in (GaAl)As diode lasers due to electron number fluctuations", *Appl. Phys. Lett.*, Vol. 40 (1982) pp. 560–562.

Chapter 5

Principles and Methods of FM Noise Reduction in Lasers

General discussions on principles, methods and techniques of FM noise reduction in lasers are presented in this chapter. Five methods are presented to reduce FM noise, which can be, in principle, employed for types of lasers.

5.1 Principles of FM Noise Reduction

It may be easily found from Eq.(4-60) that there are several methods for FM noise reduction. They are:

[1] Reduction of Γ_{ex} by reducing the magnitudes of external noise, e.g. current fluctuations from the current source, ambient temperature fluctuations and mechanical and acoustic vibrations. In the case of semiconductor lasers, one can easily reach the quantum noise level given by the first term of Eq.(4-59) because that level is fairly high.

[2] After the quantum noise level has been reached, spontaneous emission fluctuations Γ_s are expected to be reduced. For this purpose, the technique of cavity QED (quantum electrodynamics) may be employed in the near future.

[3] Reduction of the α-parameter. For this purpose, it is effective to introduce a quantum well structure. In this case, the α-parameter (see Eq.(4-56)) can be reduced to zero because the center frequencies of the spectral profiles of χ' and χ'' coincide with each other due to a modification in energy band structures by the quantum size effect.

[4] Reduction of the cavity loss Γ_q.

These four methods correspond to the optimum designs of device structures at the stage of laser device fabrication. In addition to these, there is also the method of active negative feedback by which the measured magnitude of FM noise is reduced. Even if laser devices are fabricated from careful designs based on methods [1] ~ [4], they are not always sufficiently stable. To overcome this difficulty, the following method should be employed:

[5] An electronic servo-control circuit for negative electrical feedback is connected to the laser device for FM noise reduction.

Method [5] has been popularly used as the most reliable method for optical application systems. The effect of the electrical feedback can be represented by

$$\delta\nu(t) = \Gamma_q\left(1+\alpha^2\right)\cdot\Gamma_s(t) + \Gamma_{ex}(t)$$

$$-\int_0^\infty h(\tau)\cdot\left\{\delta\nu(t-\tau) + \Gamma_n(t-\tau)\right\}d\tau \qquad (5-1)$$

The convolution integral represents the effect of the negative feedback whereas the other terms are the same as those in Eq.(4-60). In this integral, $h(\tau)$ is the impulse response function of the feedback loop, $\delta\nu(t-\tau)$ is the angular frequency fluctuations measured by the feedback loop and $\Gamma_n(t-\tau)$ is the magnitude of noise generated from the feedback loop. The Fourier transform of this relation is

$$F(f) = \frac{\Gamma_q\left(1+\alpha^2\right)}{1+H(f)}\cdot\Pi_s(f) + \frac{1}{1+H(f)}\cdot\Pi_{ex}(f) - \frac{H(f)}{1+H(f)}\cdot\Pi_n(f) \qquad (5-2)$$

where F, Π_s, Π_{ex}, Π_n and H represent the Fourier transforms of $\delta\nu$, Γ_s, Γ_{ex}, Γ_n and h, respectively. This equation shows that the first and second terms approach zero at an infinite gain in the feedback loop ($|H| \to \infty$), i.e. contributions from the quantum noise of the laser device and from external noises can be suppressed. The third term approaches $|\Pi_n|$, i.e. the magnitude of noise generated from the feedback loop. That is, if the total system of electrical feedback is treated as a quantum mechanical system, the source of quantum noise in this feedback system is not only spontaneous emission fluctuations generated in the laser cavity but also fluctuations represented by Π_n. The most fundamental contribution to Π_n is the shot noise generated from the photo-detector used as the initial step in the FM noise detection system, i.e. fluctuations of the number of electrons emitted from the photo-detector as a result of conversion from the optical lightwave to electrons. Therefore, even though the magnitude of FM noise in the laser device is large, it can be reduced to the shot noise limit by employing a high-gain electrical feedback loop [1]. If a low-noise feedback loop is employed, the magnitude of the reduced FM noise can be lowered far below that from the contribution of spontaneous emission fluctuations; this was shown in Fig. 4-10. Furthermore, the half linewidth of the field spectrum can become much narrower than the Schawlow-Townes limit in Eq.(4-52). Since this reduced magnitude of FM noise is lower than that in the coherent state of light, this state can be called a hyper-coherent state [2].

Since method [5] has been popularly used, its details are described in the following paragraphs. Subsequently, methods [2] and [4] are also described.

5.2 Negative Electrical Feedback Method

Figure 5-1 shows the block diagram of a negative electrical feedback. This feedback system has a higher gain and narrower bandwidth than those in other feedback systems. The bandwidth is mainly limited by a group delay in the feedback signal propagation determined by the length of the feedback loop. This system is composed of three parts: (1) Detection of FM noise. (2) Amplification of the control signal. (3) Modulation of frequency. The details of these parts are described below.

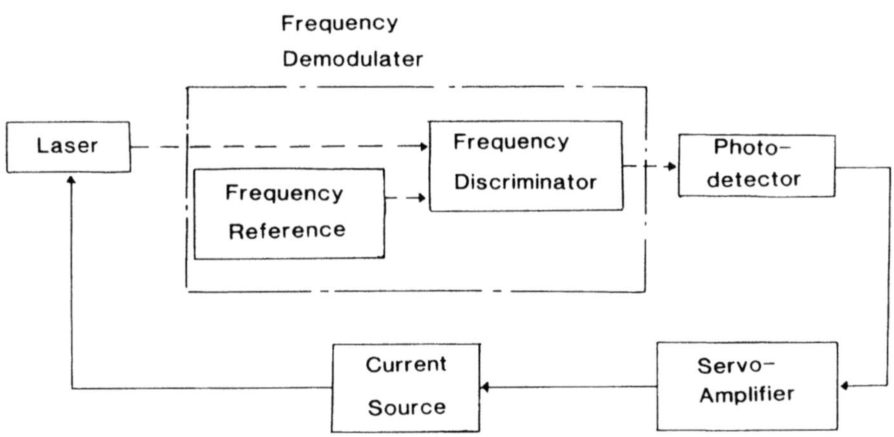

Fig. 5.1. Schematic configuration of a control system using negative electrical feedback.

5.2.1 Detection of FM noise

The frequency demodulator, composed of a frequency discriminator and a frequency reference providing the stable reference angular frequency ν_r, is required to measure the deviation in laser angular frequency, ν_{ql}, from ν_r. The resonance spectral lines of electronic or vibration-rotation transitions in stable atoms or molecules can be used as frequency references. Table 5-1 shows examples of these atoms and molecules. The performances required for these frequency references are the accurately calibrated center frequency of the spectral line and the low magnitude in frequency shift induced by several perturbations (variations in ambient electric field, magnetic field, temperature, gas pressure and so on); these two performances are required in order to guarantee accuracy and reproducibility, respectively. For the frequency discriminator, the derivative's center slope of atomic or molecular spectral profiles can be used as long as this slope can be linearly approximated (see Fig. 5.2). This derivative can be derived from phase-sensitive detection by modulating the laser frequency, for example.

Table 5.1. Atoms and molecules to be used as frequency references for laser frequency control.

Laser	Frequency Reference	Wave length(μm)
He-Ne	CH_4	3.4
	I_2	0.633
	I_2	0.612
He-Xe	H_2CO	3.5
Ar	I_2	0.515
CO	CO_2	10.6
	SF_6	10.6
Dye	I_2	0.576
	Ca	0.657
AlGaAs	H_2O	0.83
	Rb	0.78
	C_2	0.85
InGaAsP	H_2O	1.5
	NH_3	1.5
	C_2H_2	1.5
	HCN	1.5
	Kr	1.5
	Ar	1.3
InGaAlP	Li	0.67
	Ca	0.65

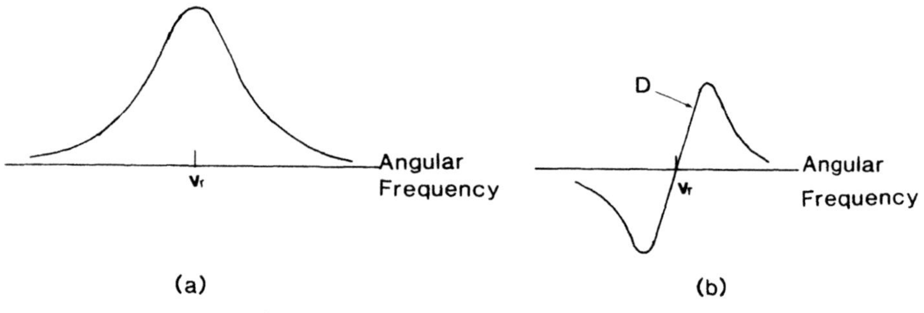

Fig. 5.2. Examples of frequency reference and frequency discriminator. (a) Spectral profile of atoms or molecules. v_r is the reference angular frequency. (b) Derivative of curve (a). D is the slope of the curve at the value of v_r.

The performances required for the frequency demodulator are low-noise, high-sensitivity and wide-bandwidth. The noise generated from the frequency demodulator could limit the accuracy of finding the center frequency of the spectral line whereas the sensitivity could limit the gain of the feedback loop. Since the sensitivity is proportional to the slope of the derivative curve in Fig. 5-2(b), it is more advantageous to use a spectral line with a narrower linewidth. The minimum detectable magnitude of FM noise is given by N/D, where D and N are the magnitude of the slope and the noise of the frequency demodulator, respectively. The quantum noise limit of the FM noise detection is given if the magnitude of N is due to shot noise in the photo-detector. If the statistical property of FM noise is represented by white noise, the one-sided power spectral density $S_{\nu\text{FB}}(f)$ and the square root of the Allan variance $\sigma_{\nu}{}^2(\tau)$ representing this limit are expressed as [3]

$$S_{\nu\text{FB}}(f) = 0\ 08(N / D)^2 \qquad (5-3)$$

and

$$\sigma_{\nu}(\tau) = \frac{0\ 2N}{\nu_r D}\tau^{-1/2} = \frac{0\ 2}{Q(S/N)}\tau^{-1/2} \qquad (5-4)$$

To derive these expressions, the slope D was approximated by $D = S/\Delta\nu_r$ and the quality factor Q of the spectral line was defined by

$$Q \equiv \nu_r / \Delta\nu_r \qquad (5-5)$$

where S is the peak-to-peak height of the derivative curve in Fig. 5-2(b) and $\Delta\nu_r$ is the FWHM of the reference spectral shape represented in the dimension of the angular frequency. The signal-to-noise ratio S/N of Eq.(5-4) represents the S/N of the spectral measurements in Fig. 5-2.

Equations (5-3) and (5-4) show that it is more advantageous to use a spectral line with a narrower linewidth, $\Delta\nu_r$, and larger intensity, S. Endeavors to develop these spectral lines have been continuously undertaken for several decades; resulting progresses are shown in Fig. 5-3 and are listed below.

(1) Linear absorption spectral line (Fig. 5-3(a)): Spectral width is determined by the first-order Doppler effect from the thermal motion of low-pressure gaseous atoms or molecules, which can be of the order of 1 GHz.

(2) Doppler-free spectral line (Fig. 5-3(b)): There are, at least, three ways of eliminating the first-order Doppler effect in order to obtain a narrow spectral linewidth (natural linewidth) which is determined by the lifetime of atomic or molecular energy levels. Such a linewidth can be narrower than a few MHz. These three approaches are: (a) Using an atomic beam. The spectral profile of an atomic beam is measured by irradiating the laser light perpendicular to the atomic beam's trajectory. (b) Using a saturated absorption spectral line. A high-power pump light is used to excite a group of atoms or molecules having a certain value of thermal

velocity in order to induce the saturated absorption of these atoms. The absorption spectral profile of these saturated atoms or molecules is selectively measured by irradiating a probe light. (c) Using a two-photon absorption spectral line. For atoms or molecules moving with a velocity component v along the $+z$ direction, the angular frequency of the irradiated running lightwave with an angular frequency ν propagating toward the +z direction is $\nu(1 - v/c)$ due to the first-order Doppler effect, where c is the speed of light. And that of the lightwave running toward the $-z$ direction is $\nu(1 + v/c)$. Therefore, the two-photon absorption process of the atom interacting simultaneously with these two running waves could cancel the Doppler-shift of $\pm \nu v/c$; as a result, a Doppler-free spectral line can be observed.

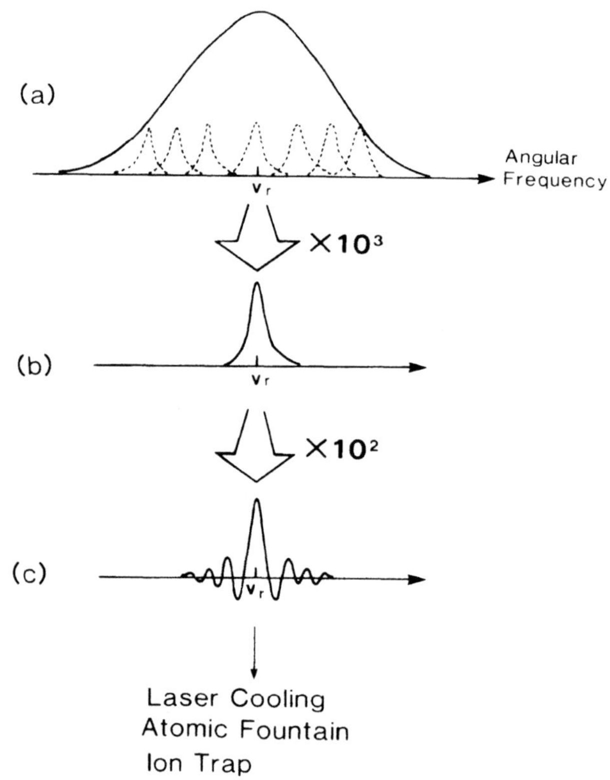

Fig. 5.3. Progresses of the spectral profiles in atoms and molecules used as the frequency reference and frequency discriminator. (a) Spectral profile of atoms or molecules with the first-order Doppler broadening. The linewidth of the profile is determined by the linewidth of the thermal velocity distribution of the atoms or molecules. (b) Doppler-free Spectral profile. Its linewidth is determined by the lifetime of each atomic energy levels. (c) Sub-natural linewidth spectral profile. The linewidth is determined by the measurement system, and the linewidth narrower than the natural linewidth is possible.

(3) Sub-natural linewidth spectral line (Fig. 5-3(c)): Even though Doppler-free spectral lines are obtained, their minimum linewidth is mainly determined by the lifetimes of the atomic or molecular energy levels and by the flight time of the atoms or molecules when traversing across the laser light beam. The former gives the natural linewidth which depends on the structures of atoms or molecules. The latter gives an the artificial linewidth which corresponds to the resolution of the spectrometer. Methods to obtain a sub-natural linewidth have been proposed. The first of those methods is the optical Ramsey method [4] obtained by irradiating three or four spatially separated parallel laser beams on the atomic beam in a direction perpendicular to the atomic beam's trajectory. In this configuration, atoms could suffer a temporally intensity-modulated perturbation from these laser beams. This perturbation could induce spectral profile modulations. This type of modulated profile is called an optical Ramsey fringe. The width of the main lobe is inversely proportional to the traversing time of the free-space between adjacent laser beams. A linewidth of 3 kHz can be obtained when the atomic thermal velocity is 300 m/s and separation between laser beams is 10 cm. This kind of modulation technique has been employed in a variety of microwave application systems to improve the resolution of measurements. For example, they have been used in side-looking radars, very long baseline microwave interferometers (VLBI), and in microwave Ramsey fringes in Cesium beam atomic frequency standards. Although this is a technique which is already established in the microwave frequency region, there were several experimental difficulties in the optical frequency region because of the short optical wavelength. Fortunately, this technical difficulty has been overcome and optical Ramsey spectral lines having a sub-natural linewidth have been successfully obtained [5].

The second method uses the spectral profile from atoms whose lifetimes are longer than that of the ensemble average in the atomic beam. This profile has been obtained by modulating the laser beam phase irradiated on the atomic beam [6].

Methods (1) and (2) have been employed to obtain a reliable frequency demodulator for laser frequency control. Method (3) is still under development. Although the following three methods, (4)-(6), are still at a preliminary stage of study, their progress is very striking and further developments are expected:

(4) Cooling and confinement of atoms [7].

(5) Atomic fountain [8].

(6) Ion trap [9].

The details of these methods will be described in 7.5(3).

Since the spectral profiles in Fig. 5-2(a) obtained by methods (1) ~ (6) cannot be used directly, their derivatives corresponding to the curve in Fig. 5-2(b) should be obtained by frequency modulation and phase-sensitive detection. However, a spectral profile similar to that in Fig. 5-2(b) can be alternatively used if the frequency dependence of the refractive index, i.e., the dispersion characteristics, of the atomic or molecular medium is obtained by measuring the change of phase in laser light transmitted through this medium. These dispersion characteristics correspond to the real part of the complex susceptibility χ' in the medium. Meanwhile, the profile of the curve in Fig. 5-2(a) corresponds to the frequency dependence of the imaginary part of the complex susceptibility χ''. Although the profile of χ' is similar to that of the curve in Fig. 5-2(b), it is not the exact derivative of the curve in Fig. 5-2(a). Frequency modulation is not required if this profile is used as a frequency demodulator. An example of measuring χ' is shown in Fig. 5-4, where the profile of dispersion of the saturated absorption in CH_4 is given [10].

Spectral lines in stable atoms or molecules could provide a high-quality reference angular frequency ν_r and its value has been calibrated based on the study of atomic or molecular structure. However, if the absolute value of ν_r is not required in some application systems, a simpler frequency demodulator can be used. One such example is the Fabry-Perot interferometer, or the so-called Fabry-Perot etalon described in 3.2. Although this is exactly the same as that shown in Fig. 2-1(b), it is, however, used not as a laser cavity but as a frequency demodulator for the present purpose. Its resonance characteristics are shown in Fig. 2-4(b), in which the angular frequency of one longitudinal mode works as the reference angular frequency ν_r.

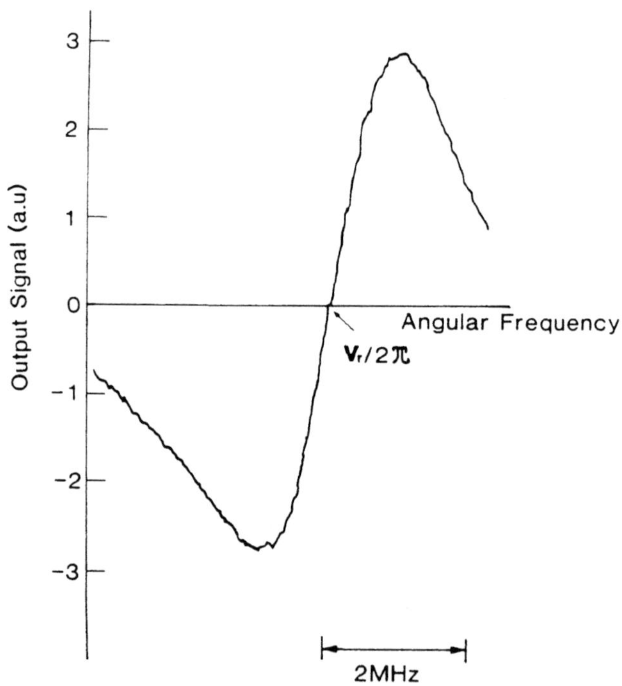

Fig. 5.4. Measured result of a saturated dispersion spectral shape of CH_4 (after Ohtsu, Ohta, and Tako [10]).

Although atomic or molecular spectral profiles are sometimes complicated because of their hyperfine structures, the resonance curve of the Fabry-Perot interferometer is simpler, therefore, it can work as a high quality frequency reference if the values n_q and L of Eq.(2-7) are stable. Figure 5-5(a) shows the magnified profile of one of the resonance curves in Fig. 2-4(b). This profile can be obtained by measuring the light power transmitted through the Fabry-Perot interferometer as a function of light frequency. The derivative of this curve, as shown in Fig. 5-5(b), can be obtained by modulating the laser frequency and can be used as frequency demodulator. The slope part of the curve in Fig. 5-5(a) can also be used as frequency demodulator by offsetting the output signal of the photo-detector used in measuring transmitted light power. In this case, the laser frequency does not have to be modulated; however, the reference frequency is not ν_r any longer. Since the sensitivity of the frequency demodulator is proportional to the slope of the curve in Fig. 5-5(a) or (b), it is more advantageous to use a Fabry-Perot interferometer with a narrow resonance linewidth ($\Delta\nu_c$ of Eq.(2-15)). That is, the gain G of the frequency demodulator is inversely proportional to $\Delta\nu_c$. In the field of optics, a measure is used to represent the narrowness of this resonance curve, which is called the finesse and is defined by

$$F \equiv \frac{\Delta\nu_q}{\Delta\nu_c} = \frac{\pi\sqrt{R}}{1-R} \qquad (5-6)$$

where the internal loss of the interferometer (α_l of Eq.(2-10)) is neglected and the power reflectivities of the two mirrors are assumed as R. Although this measure F

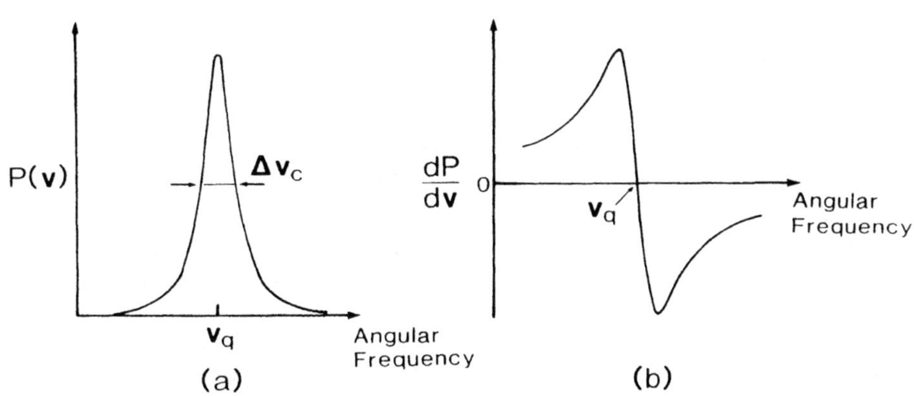

Fig. 5.5. Spectral profile of the q-th longitudinal mode of a Fabry-Perot interferometer. P is the transmitted light power, ν is the angular frequency of incident light. (a) Resonance spectral profile with the resonance angular frequency of ν_q. The linewidth is determined by Eqs.(2.7) and (2.15). (b) The derivative of the curve in (a).

is almost equivalent to the quality factor in Eq.(2-16), the longitudinal mode separation $\Delta\nu_q$ (this is called the free spectral range (FSR) in the field of optics) appears in the nominator of Eq.(5-6) instead of the resonance angular frequency ν_q. A super-cavity with a finesse as high as 1,000,000 has been developed. Since the resonance linewidth $\Delta\nu_c$ of this super-cavity can be narrower than 100 kHz, i.e. even narrower than the saturated absorption linewidth of atoms or molecules, it can be used as a highly sensitive frequency demodulator. However, it should be noted that very fast FM noise components cannot be measured by these highly sensitive frequency demodulators. That is, the bandwidth B of this measurement is proportional to $\Delta\nu_c$, which means that the gain G and bandwidth B are inversely proportional to each other. This relation holds also for the atomic or molecular spectral lines in (1) ~ (6), which is equivalent to the fact that the gain-bandwidth product, GB, is a constant in the electronic amplifier.

To discuss the bandwidth, one has to investigate the transfer function of the Fabry-Perot interferometer $H_{\mathrm{FP}}(f_m)$, which is defined by

$$H_{\mathrm{FP}}(f_m) \equiv \partial P(f_m) / \partial \nu_l(f_m) \tag{5-7}$$

where P is the output laser power from the Fabry-Perot interferometer, ν_l is the angular frequency of the laser incident onto the Fabry-Perot interferometer and f_m is the modulation frequency of the laser frequency. Although Fig. 5-5(a) shows the value of P which is measured for $f_m = 0$, i.e. by sweeping ν_l adiabatically, the dynamic response for a higher modulation frequency should be known for the present discussion. Since the output $P(t)$ has a time-delay to the input $\nu_l(t)$, the transfer function $H_{\mathrm{FP}}(f_m)$ usually takes on a complex value.

Figures 5-6(a) and (b) show the absolute value and phase of the transfer function obtained by numerical calculations [11, 12]. Curve A is the result by measuring the laser power transmitted through the Fabry-Perot interferometer, which is called the transmission-mode. Curve A in (a) shows that its cut-off frequency is $(\Delta\nu_c/2)/2\pi$. Since the gain $|H_{\mathrm{FP}}(0)|$ is inversely proportional to $\Delta\nu_c$, it can be confirmed that the GB product is independent of $\Delta\nu_c$. In the range of $f_m > (\Delta\nu_c/2)/2\pi$, the value of curve A exhibits a -40 dB/decade decrease proportional to f_m^{-2}. Curve A in (b) shows that the phase changes 180 degree at the limit $f_m \rightarrow \infty$. On the other hand, curve B in (a) represents the result obtained by measuring the power reflected from the Fabry-Perot interferometer, which is called the reflection-mode. Although the cut-off frequency is the same as that in curve A, it exhibits a -20 dB/decade decrease at $f_m > (\Delta\nu_c/2)/2\pi$, i.e., $|H_{\mathrm{FP}}(f_m)|$ is proportional to f_m^{-1}. Since the slope of this curve is smaller than that of curve A, a wider bandwidth of frequency demodulation is expected. Furthermore, curve B in (b) shows that the phase changes only 90 degree at $f_m \rightarrow \infty$, which is more advantageous than the transmission-mode to realize a wider bandwidth of negative electrical feedback. The difference between the transmission- and reflection-modes can be interpreted

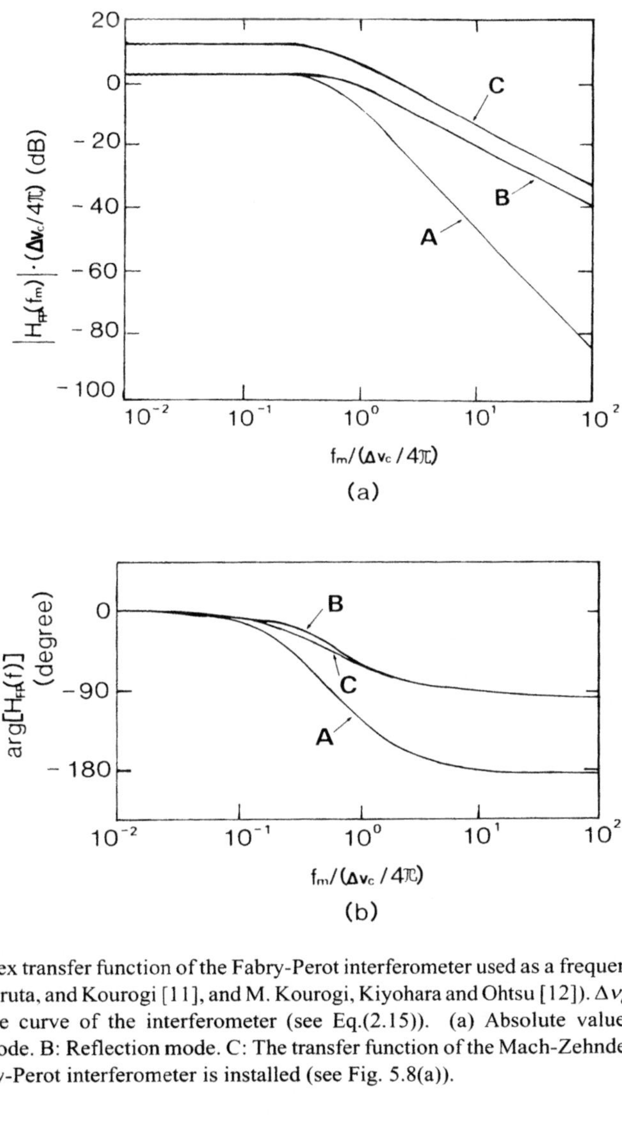

Fig. 5.6. Complex transfer function of the Fabry-Perot interferometer used as a frequency demodulator (after Ohtsu, Muruta, and Kourogi [11], and M. Kourogi, Kiyohara and Ohtsu [12]). $\Delta\nu_c$ is the linewidth of the resonance curve of the interferometer (see Eq.(2.15)). (a) Absolute value. (b) Phase. A: Transmission mode. B: Reflection mode. C: The transfer function of the Mach-Zehnder interferometer in which a Fabry-Perot interferometer is installed (see Fig. 5.8(a)).

as follows; as is shown in Fig. 5-7, reflected light contains two kinds of lights, the high-intensity directly reflected light from the input mirror of the Fabry-Perot interferometer and light transmitted after resonance inside the Fabry-Perot interferometer. Since the incident optical frequency varies nonadiabatically within a range of $f_m > (\Delta\nu_c/2)/2\pi$, the former bears the present optical angular frequency $\nu_l(t)$ while the latter bears the past averaged optical angular frequency $\nu_l(t-\tau)$ due to the resonance inside the interferometer, where τ is the averaged time required to travel

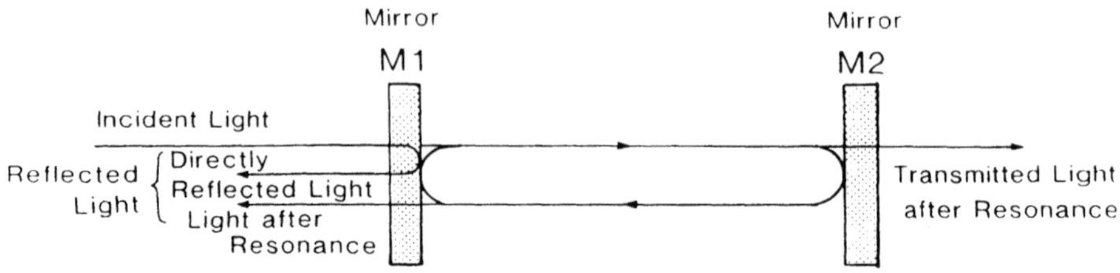

Fig. 5.7. Schematic explanation for the property of the transfer function of transmission and reflection modes of a FP interferometer.

inside the interferometer. Therefore, measurement of the reflection mode corresponds to measuring the difference between $v_l(t)$ and $v_l(t-\tau)$, which can be approximated as a temporal differentiation, i.e.

$$v_l(t)-v_l(t-\tau)\propto \mathrm{d}v_l(t)/\mathrm{d}t \qquad (5-8)$$

The differentiation operator in this equation corresponds to $i(2\pi f)$ in the Fourier frequency domain. On the other hand, the transmission mode does not contain directly reflected light from the input mirror which serves as reference for the differentiation. Therefore, due to the effect of this differentiation operator, the slope and the maximum phase change of the reflection-mode are +20dB/decade larger and 90 degree smaller than those of the transmission mode, respectively. Owing to these reasons, a wider bandwidth frequency demodulator can be realized by using reflection-mode.

One of the principal noises contained in the output signal from these frequency demodulators is the IM noise of light. The shot noise limit is reached by eliminating the contribution from IM noise. Figures 5-8(a) and (b) show two methods proposed to achieve the shot noise limit; a Mach-Zehnder interferometer in which a Fabry-Perot interferometer is installed (MZ/FP interferometer) [12] and a Fabry-Perot interferometer in which a linear polarizer is installed [13]. In (a), a Mach-Zehnder interferometer is used to realize the interference between transmitted and reflected light from the Fabry-Perot interferometer. The output from this MZ/FP interferometer shows specific characteristics caused by the difference in dynamic properties between transmitted and reflected light from the Fabry-Perot interferometer. If the length of the Fabry-Perot interferometer L_{FP} is equal to the difference between the two arms of the Mach-Zehnder interferometer ΔL_{MZ} $(=L_1-L_2)$, the optical frequency dependence of the output signal from the photo-detector D1 of Fig. 5-8(a) gives a dispersive lineshape, as is shown in Fig. 5-9(a). Since the signal from the photo-detector D2 takes an opposite sign for this profile, the output signal from the differential amplifier gives the curve shown in Fig. 5-9(b). In this case, the two photo-detectors and the differential amplifier work together as an optical balanced detector and the common mode noise due to the IM noise of the laser can be removed

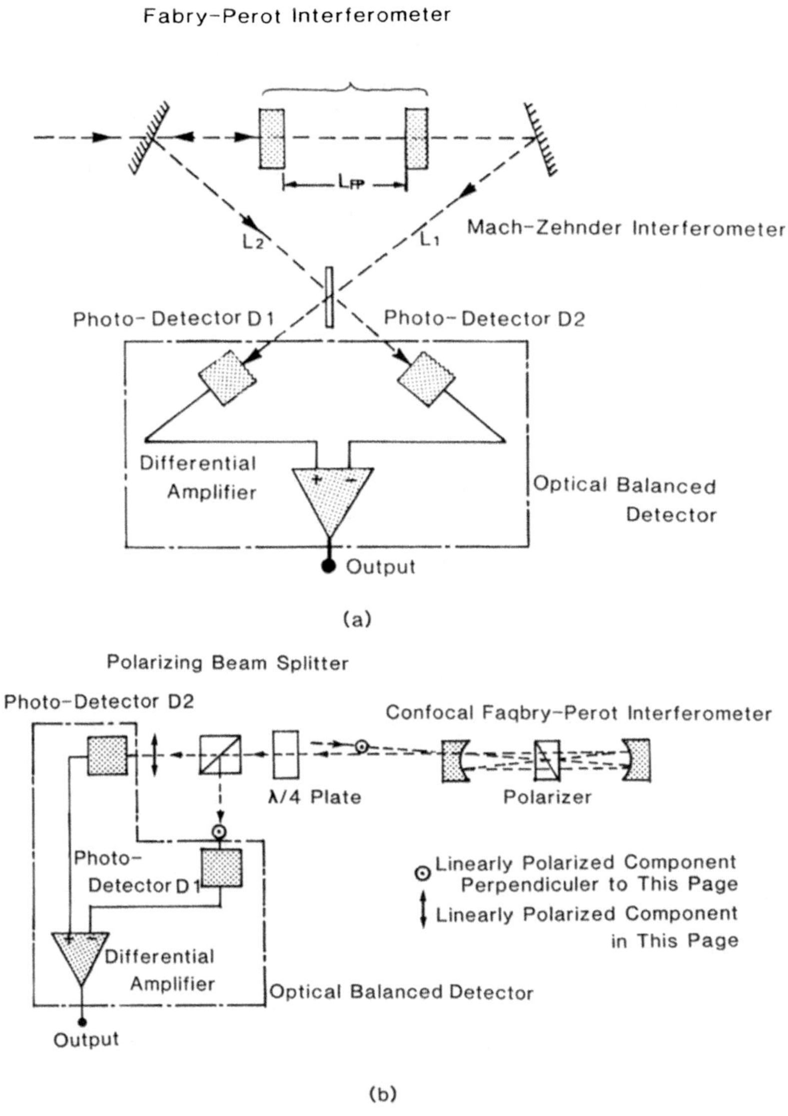

Fig. 5.8. Possible demodulator for eliminating the effect of the IM noise of a laser. (a) A Mach-Zehnder interferometer in which a FP interferometer is installed (after Kourogi, Kiyohara, and Ohtsu [12]). L_{FP} is the length of a Fabry-Perot interferometer. L_1 and L_2 are the lengths of the two arms of the Mach-Zehnder interferometer. (b) A Fabry-Perot interferometer in which a polarizer is installed (after Hänsch and Couillaud [13]).

by the differential amplifier; therefore, the optical balanced detector can reach the shot noise limit. Furthermore, this setup can be advantageously used as a frequency demodulator because the dispersive shape of Fig. 5-9(b) is obtained without modulating the laser frequency. The transfer function of this frequency demodulator is given by curve C in Fig. 5-6. It is found that the gain is 10 dB larger than that of the reflection mode of the Fabry-Perot interferometer and that the phase change is only 90 degree, which means that it is more advantageous than the reflection-mode of a Fabry-Perot interferometer. In Fig. 5-8(b), a linearly polarized compo-

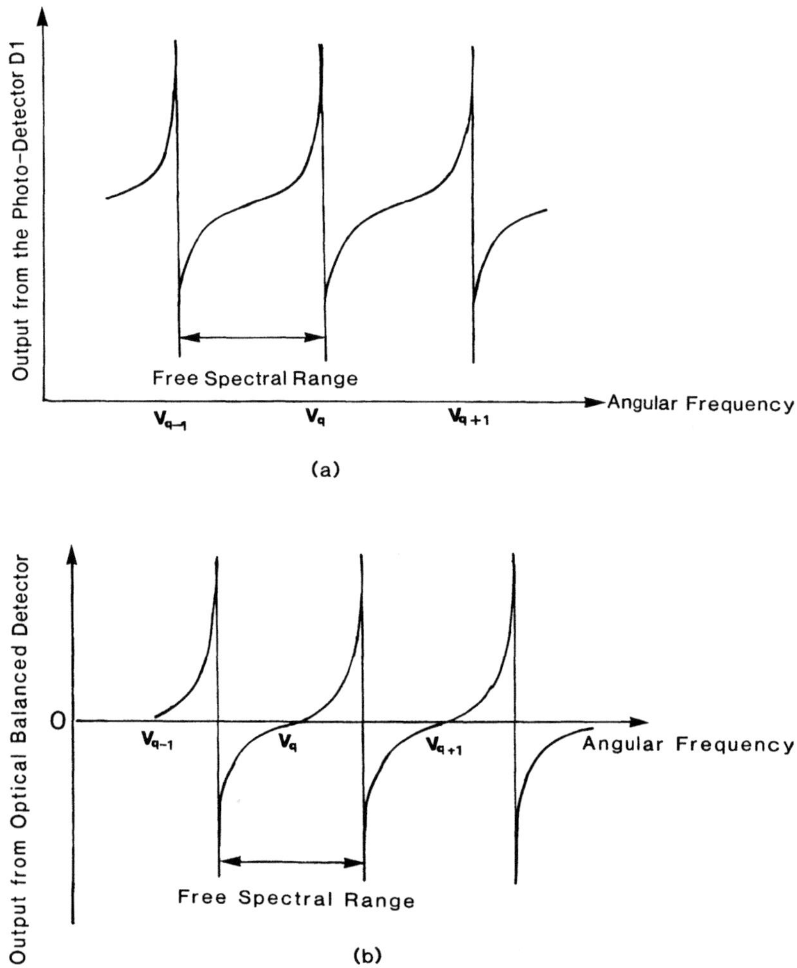

Fig. 5.9. Frequency demodulation charateristics of the setup in Fig. 5.8. (a) Output of the detector D1 in Fig. 5.8(a). (b) Output of the optical balanced detectors in Figs. 5.8(a) and (b).

nent parallel to the optical axis of the linear polarizer can resonate in the Fabry-Perot interferometer, while the perpendicular component is directly reflected from the input mirror. The light reflected back from the Fabry-Perot interferometer is composed of these two components. Therefore, the dispersive lineshape, as is shown in Fig. 5-9(b), is obtained by measuring the change in polarization of the light by the optical balanced detector composed of D1, D2 and a differential amplifier. The effect of the IM noise can also be removed to achieve the shot noise limit.

The disadvantage of the MZ/FP interferometer, Fig. 5-8(a), is that a change in L_{FP} or ΔL_{MZ} could lead to the deformation of the dispersive lineshape of Fig. 5-9, which could deteriorates the performance of the frequency demodulator. To overcome this difficulty, a more stable frequency demodulator has been proposed [14], as is shown in Fig. 5-10(a), in which a quarter-wave plate is installed in a Fabry-Perot interferometer, for which the operation principle and its performance are almost the same as those of the MZ/FP interferometer. This Fabry-Perot interferometer with a quarter-wave plate can be replaced by a Fresnel rhomb, where both ends of the Fresnel rhomb are high-reflectively coated(Fig. 5-10(b)). It can also be replaced by the ring-resonance configuration of a confocal Fabry-Perot interferometer in which a $\lambda/8$ plate has been installed (Fig. 5-10(c)). Fig. 5-10(d) shows the measured dispersive lineshape. Furthermore, as is shown by Figs. 5-10(e) and (f), the ring cavity-type and the fiber ring cavity-type are also proposed as a modified setup of Fig. 5-10(a) [14]. The advantage of the ring cavity configuration is that the gain of frequency discrimination can be increased by 12 dB because any additional beam splitters are not required. The residual magnitude of FM noise in the controlled laser is expressed by Eq.(5-3) or (5-4) and it represents the frequency stability. However, frequency accuracy and reproducibility are also important in practical applications. To improve these, it should be noted that the reference angular frequency ν_r usually depends on environmental changes, e.g. changes in temperature, stray electromagnetic field, air pressure, mechanical vibration and acoustic vibration. Furthermore, it can also depend on the wavefront and the spot size of the incident laser light. On the other hand, since the vibration-rotation spectral line of a symmetric molecule, e.g., CH_4, is insensitive to these changes, it can be used as a stable frequency reference. Spectral lines due to electronic transitions in atoms can sometimes be frequency-shifted by the ambient stray electromagnetic field. Frequency shift of the Fabry-Perot interferometer is induced by changes in the refractive index of the medium in the interferometer and by the separation between mirrors. However, they can be reduced by using a low expansion coefficient material (e.g., Zerodur, mixture of ceramics and glass made by Schott Optical Glass Inc.) as a spacer to fix the separation between the two mirrors, by suspending the Fabry-Perot interferometer in a vacuum chamber and by installing an acoustic shield. Furthermore, the use of glass made of SiO_2 and TiO_2 (ULE made by Corning Glass Works, with an expansion coefficient of 1×10^{-9}/degree) has been proposed [15].

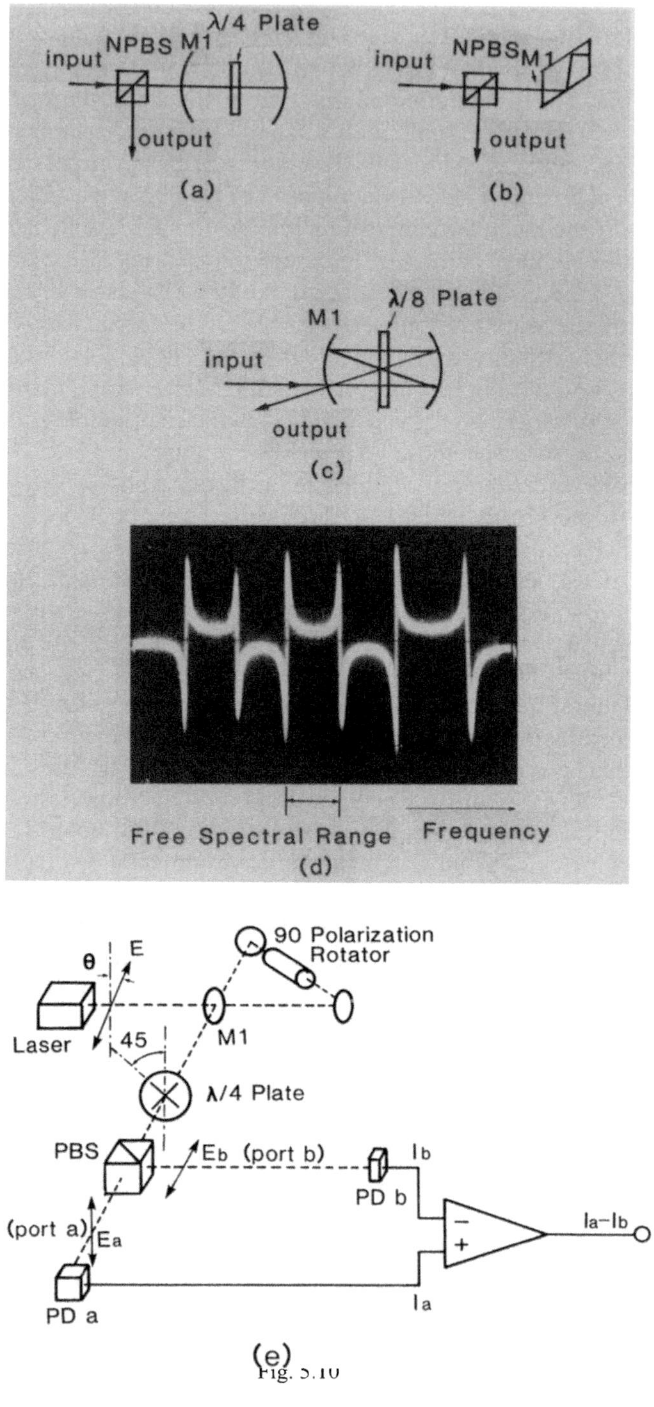
λ/4 Plate
NPBS
M1
input
output
(a)
input
NPBS
M1
output
(b)
λ/8 Plate
M1
input
output
(c)
Free Spectral Range
Frequency
(d)
90 Polarization Rotator
θ
E
Laser
45
M1
λ/4 Plate
PBS
Eb (port b)
Ib
PD b
(port a)
Ea
PD a
Ia
Ia−Ib
−
+
(e)

Fig. 5.10

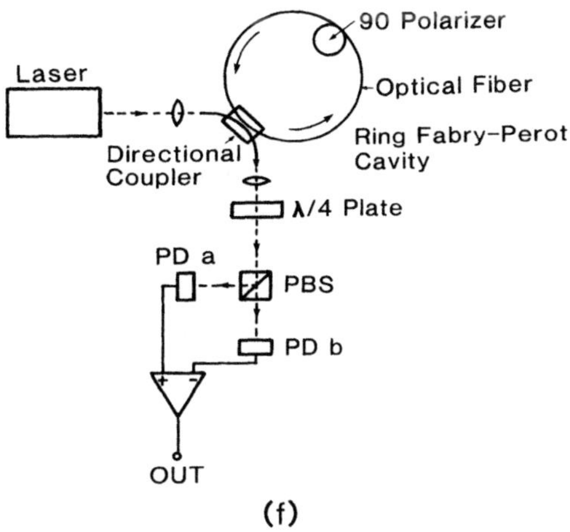

Fig. 5.10. Modified frequency demodulator for more stable operation (after Kougori and Ohtsu [14]). (a) A Fabry-Perot interferometer in which a quarter-wave plate is installed. NPBS: Non-polarized beam splitter. (b) A Fresnel rhomb, whose two facets are high-reflection coated. (c) A ring-resonance configuration of a confocal Fabry-Perot interferometer in which a λ/8 plate is installed. (d) Measured dispersive lineshape of the demodulator. (e) Ring cavity-type. (f) Fiber ring cavity-type.

5.2.2 Amplification of the control signal and modulation of frequency

The error signal from a frequency demodulator, being proportional to $\nu_l - \nu_r$, is amplified by servo-amplifier and its output signal is applied to the frequency modulator. This amplifier can be optimally designed to maximize the gain and bandwidths of the loop transfer function by computer simulation based on the transfer function of each element in the control loop. This design criterion has been conventionally used in analog electronic circuits and its high accuracy, reliability and reproducibility have been confirmed. Because of these advantages, negative electronic feedback has been evaluated as the most reliable method.

The design of the servo-amplifier depends on the property of a frequency modulator, i.e. on the transfer function representing the laser frequency shift by unit change of input signal into the modulator. In the case of semiconductor lasers, for example, it corresponds to the frequency modulation efficiency of the injection current, as shown in Fig. 3-30. For the frequency modulations of other lasers, a piezo-electric transducer (PZT), a phase modulator using the electro-optical effect (E/O effect) and a frequency shifter using the acousto-optical effect (A/O effect) can be employed because the modulation efficiency by modulated population inversion is much lower than that in a semiconductor laser. Although the response speed of the PZT is usually low, a bandwidth as wide as 1 MHz can be achieved by reducing the size of the PZT and by burying the PZT into the PbO gel to absorb the acoustic wave generated from the PZT. The bandwidths of a bulk-type E/O and A/O modulators can be several MHz, while those of a waveguide-type can be as wide as several GHz.

5.3 Cavity Loss Reduction Method

This method has been mainly employed for semiconductor lasers because their cavity losses are large. One example is optical feedback [16] which is based on the phenomenon of injection locking as shown in Fig. 5-11. Injection locking is a nonlinear phenomena which has been found in general self-sustained oscillators. Its characteristics can be analyzed by using Van der Pol's equation in Eq.(2-83) [17], i.e. the slave laser frequency is pulled to that of the master laser by injecting the master laser light into the slave laser cavity. The locking range, i.e. the maximum frequency detuning between the two lasers for the realization of the injection locking, is given by [18]

$$-\frac{1}{2\tau_p}\sqrt{\frac{P_m}{P_s}} \leq \nu_m - \nu_s \leq \frac{1}{2\tau_p}\sqrt{\frac{P_m}{P_s}\left(1+\alpha^2\right)} \qquad (5-9)$$

where P_m, P_s, ν_r and ν_s are the powers and the angular frequencies of the master and slave lasers, respectively. One should note that P_m and P_s represent the values measured inside the slave laser cavity. τ_p is the photon lifetime of the slave laser and α is the α-parameter of the slave laser. Since this equation shows that the locking range is inversely proportional to the photon lifetime, i.e. proportional to the magnitude of the cavity loss, semiconductor lasers can have a wide locking range and be injection-locked easily. Figure 5-12 shows the measured result of a locking range in a semiconductor laser [19]. FM noise in the slave laser can be reduced by injection locking from a stable master laser. The locking range corresponds to the bandwidth of this FM noise reduction. The slave laser frequency can be swept in a stable manner by sweeping that of the master laser within the locking range. However, the accuracy and reproducibility of the frequency sweep are usually lower than those in negative electrical feedback because injection locking is based on the nonlinearity of oscillators.

Optical feedback corresponds to injection locking by injecting the light reflected back from an external reflection surface into the laser cavity. Although deterministic instabilities are usually induced by injecting this reflected light (see

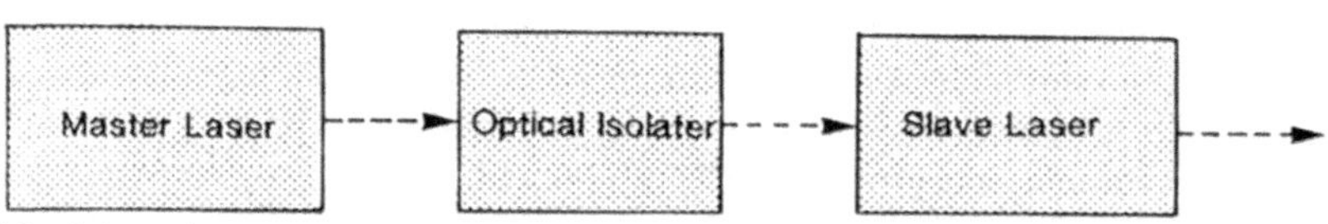

Fig. 5.11. Schematic explanation of an injection locking experiment. An optical isolator is used for preventing the light injection from slave laser to master laser.

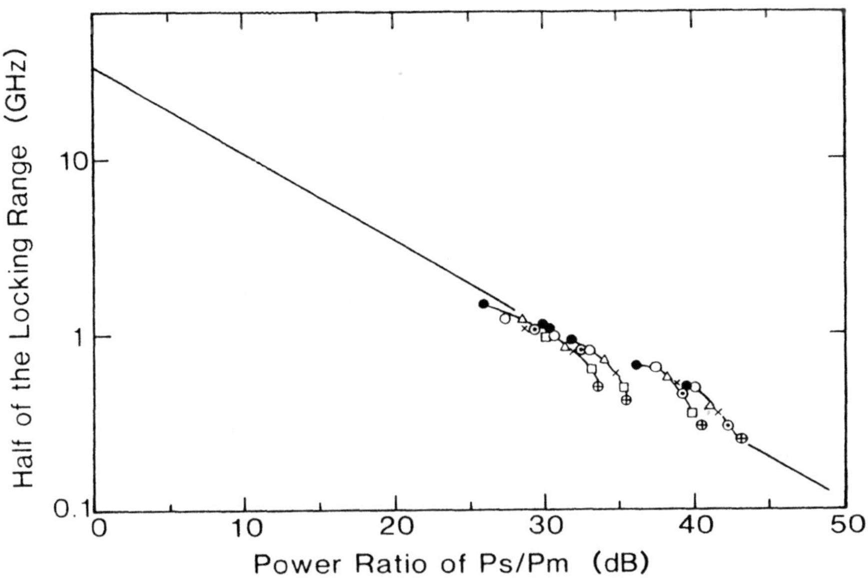

Fig. 5.12. Locking range of an injection locked semiconductor laser (after Kobayashi and Kimura [19], ©1981 IEEE). P_m and P_s are the powers of the master and slave lasers, respectively. The solid line is a calculated result.

4.3.2), a stable optical feedback can be achieved if the phase of the reflected light is equal to that of the light emitted from the laser cavity (i.e. when the separation between the laser cavity and the external reflector is an integer multiple of the half wavelength) and when the reflectivity of the cavity output mirror is sufficiently low to ensure a sufficient power of reflected light. These conditions converge to form a large laser cavity by using an external reflector (i.e. an external cavity) and to reduce the cavity loss. This method is effective especially for semiconductor lasers and has been used for over 15 years. However, it should be called an opto-mechanical feedback because electro-mechanical elements such as PZT have been used to control the separation between laser and external reflector. For optically feedback semiconductor lasers, frequency modulation efficiencies are reduced drastically because the photon lifetime determined by the external cavity becomes very long. Furthermore, hopping occurs between external cavity modes because the separation between these modes is far narrower than the gain spectral linewidth. The optical feedback shows a narrow bandwidth because it is inversely proportional to the photon lifetime of the external cavity.

An advanced semiconductor laser device employing this optical feedback method has been fabricated by integrating an external reflector, passive waveguide and laser on a common substrate [20]. In this device, the feedback bandwidth is wide because the photon lifetime is not very long. The gain of this feedback is low-to-

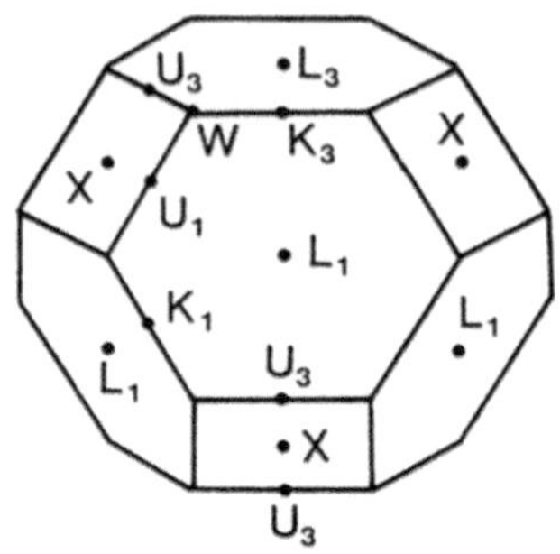
U3
L3
W
K3
X
X
U1
L1
K1
L1
L1
U3
X
U3

(a)

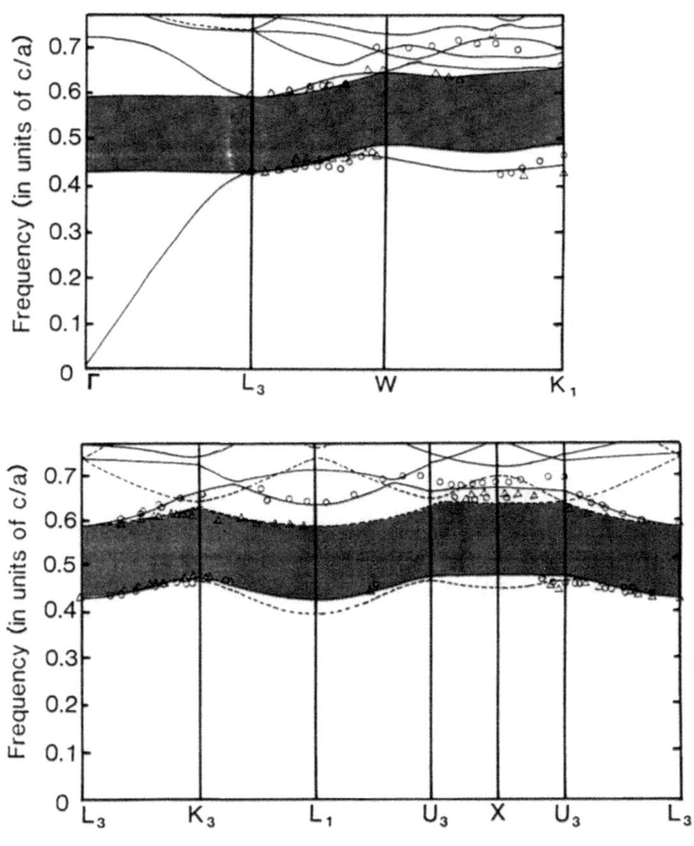
Frequency (in units of c/a)
0.7
0.6
0.5
0.4
0.3
0.2
0.1
0
Γ
L3
W
K1
Frequency (in units of c/a)
0.7
0.6
0.5
0.4
0.3
0.2
0.1
0
L3
K3
L1
U3
X
U3
L3

(b)

Fig. 5.13

medium. Another method, mainly employed for semiconductor lasers, is to reduce the cavity loss by reducing the internal cavity loss α_l at the stage of laser fabrication, by increasing the cavity length to about 1 mm (about three times longer than in common devices) and by coating a high-reflection film on the laser cavity facet. In this case, the gain of the FM noise reduction is low-to-medium and the bandwidth can be wide.

5.4 Inhibition and Enhancement of Spontaneous Emission

Since the spontaneous emission energy is proportional to the number of cavity modes (see Eqs.(2-35), (2-37), (2-39) and (2-40)), the spontaneous emission rate can be decreased by confining the active medium in the micro-cavity. On the other hand, it can be increased by using a highly mode-degenerated cavity. This method of inhibiting or enhancing the spontaneous emission is called cavity quantum electrodynamics (cavity QED). Progresses in cavity QED have rapidly developed in recent years [21]. Experiments have been carried out in the microwave frequency region at the initial stage of study [22] and have also become possible in the optical frequency region owing to improvements in fabrication techniques of the micro-cavity [23]. The control of spontaneous emission has been demonstrated in the optical frequency region by using a high-Q and mode-degenerated Fabry-Perot cavity [24]. As an alternative method, the reduction of spontaneous emission by photon localization has been proposed [25, 26, 27] by periodically modulating the refractive index of the medium to confine light in order to realize a band gap in the photon energy level structure. An example of this band gap is shown in Fig. 5-13 [27]. This phenomenon is analogous to the Anderson localization of the electron wavefunction in the random medium [28]. Although they are still at preliminary stage of study, it can be expected that, in the future, the inhibition of spontaneous emission could be used in a low noise laser device, while the enhancement of spontaneous emission could be used in a low threshold laser device.

Fig. 5.13. An example of the band gap of photon energy (after Yablonovitch, Gmitter, and Leung [27]). (a) The Brillouin Zone of a face-centered-cubic (fcc) structure incorporating non-spherical atmos. Since the space lattice is not distorted, this is simply the standard fcc Brillouin Zone lying on a hexagonal face rather than the usual cubic face. Only the L-points on the top and bottom hexagons are 3-fold symmetry axes. Therefore they are labeled L_3. The L-points on the other 6 hexagons are labeled L_1. The U_3-K_3 points are equivalent since they are a reciprocal lattice vector apart. Likewise the U_1-K_1 points are equivalent. (b) Frequency versus wave-vector, dispersion along the surface of the Brillouin Zone shown in (a), where *c/a* is the speed of light divided by the fcc cube length. The circles and triangles are the experimental points for s and p polarization respectively. The dark shaded band is the totally forbidden bandgap. The lighter shaded stripes above and below the dark band are forbidden only for s and p polarization respectively.

[References]

1) M. Ohtsu, "Realization of Ultrahigh Coherence in Semiconductor Lasers by Negative Electrical Feedback", *J. Lightwave Technol.*, Vol. 6 (1988) pp. 245–256.
2) M. Ohtsu, "Realization of Hyper-Coherent Light", *Science*, March (1989) pp. 64–73 (in Japanese).
3) P. Kartaschof, *Frequency and Time,* Academic Press, London, 1978.
4) K. Shimoda, "Side-looking radars and optical Ramsey resonances", *Butsuri* (Monthly Journal of the Physical Society of Japan), Vol. 36 (1981) pp. 590–598 (in Japanese).
5) J. Helmcke, J. Ishikawa, and F. Riehle, "High-Contrast, High-Resolution Single Recoil Component Ramsey Fringes in Ca", *Frequency Standards and Metrology* (Proc. 4-th Symposium, Ancona, Italy, September 5-9, 1988), Ed. by A. De Marchi (Springer, Berlin, 1989) pp. 270–281.
6) F. Shimizu, K. Shimizu, and H. Takuma, "Subnatural-linewidth laser spectroscopy by phase switching", *Phys. Rev. A*, Vol. 28 (1983) pp. 2248–2253.
7) W. D. Phillips, J. V. Prodan and H. J. Metcalf, "Laser cooling and electromagnetic trapping of neutral atoms", *J. Opt. Soc. Am. B*, Vol. 2 (1985) pp. 1751–1767.
8) R. G. Beausoleil and T. W. Hansch, "Two-photon optical Ramsey spectroscopy of freely falling atoms", *Opt. Lett.*, Vol. 10 (1985) pp. 547–549.
9) W. M. Itano, J. C. Bergquist and D. J. Wineland, "Laser Spectroscopy of Trapped Atomic Ions", *Science*, Vol. 237 (1987) pp. 612–617.
10) M. Ohtsu, S. Ohta, and T. Tako, "Frequency Stabilization of a He-^{22}Ne Laser by Intracavity Polarization Spectroscopy of CH_4", *Jpn. J. Appl. Phys.*, Vol. 20 (1981) pp. 1701–1707.
11) M. Ohtsu, M. Murata, and M. Kourogi, "FM Noise Reduction and Subkilohertz Linewidth of an AlGaAs Laser by Negative Electrical Feedback", *IEEE J. Quantum Electron.*, Vol. 26 (1990) pp. 231–241.
12) M. Kourogi, A. Kiyohara, and M. Ohtsu, "Improvement of the Electrical Negative Feedback System for FM Noise Reduction of a Semiconductor Laser", *IEICE Technical Report*, Vol. 89 (1989) paper No. OQE89-62 (The Inst. Electron., Information, and Commun. Eng, Tokyo, 1989) (in Japanese).
13) T. W. Hänsch and B. Couillaud, "Laser frequency stabilization by polarization spectroscopy of a reflecting reference cavity", *Opt. Commun.*, Vol. 35 (1980) pp. 441–444.
14) M. Kourogi and M. Ohtsu, "Novel optical frequency discriminator for FM noise reduction of semiconductor lasers", Opt. Commun., Vol. 81 (1991), pp. 204–208.
15) D. Hils and J. L. Hall, "Ultra-Stable Cavity-Stabilized Lasers with Subhertz Linewidth", *Frequency Standards and Metrology* (*Proceedings of the 4-th Symposium, Ancona, Italy, Sept.5-9, 1988*) Ed. by A. De Marchi (Springer, Berlin, 1989) pp. 162–173.
16) A. Mooradian, "High Resolution Tunable Infrared Lasers", in *Laser Spectroscopy*, Ed. by R. G. Brewer and A. Mooradian (Plenum Press, New York, 1973) pp. 223–236.
17) K. Shimoda, T. Yajima, Y. Ueda, and T. Kasuya, Quantum Electronics, Syokabo Publishing, Tokyo, 1972 (in Japanese).
18) R. Lang, "Injection Locking Properties of a Semiconductor Laser", *IEEE J. Quantum Electron.*, Vol. QE-18 (1982) pp. 976–983.
19) S. Kobayashi and T. Kimura, "Injection Locking in AlGaAs Semiconductor Laser", *IEEE J. Quantum Electron.*, Vol. QE-17 (1981) pp. 681–689.
20) T. Fujita, J. Ohya, K. Matsuda, M. Ishino, H. Sato, and H. Serizawa, "Narrow spectral linewidth characteristics of monolithic integrated-passive-cavity InGaAsP/InP semiconductor lasers", *Electron. Lett.*, Vol. 21 (1985) pp. 374–376.
21) S. Haroche and D. Kleppner, "Cavity Quantum Electromagnetics", *Phys. Today*, January (1989) pp. 24–30.
22) D. Meschede, H. Walther, and G. Muller, "One-atom maser", *Phys. Rev. Lett.*, Vol. 54 (1985) pp. 551–554.

23) J. L. Jewell, S. L. McCall, Y. H. Lee, A. Scherer, A. C. Gossard, and J. H. English, "Lasing characteristics of GaAs microresonators", *Appl. Phys. Lett.*, Vol. 54 (1989) pp. 1400–1402.
24) D. J. Heinzen, J. J. Childs, J. F. Thomas and M. S. Feld, "Enhanced and Inhibited Visible Spontaneous Emission by Atoms in a Confocal Resonator", *Phys. Rev. Lett.*, Vol. 58 (1987) pp. 1320–1323.
25) E. Yablonovitch, "Inhibited spontaneous emission in solid-state electronics", *Proc. the 15th International Quantum Electronics Conference* (*IQEC'87*), Baltimore 1987 (OSA, Washington, D.C., 1987) THAA4.
26) E. Yablonovitch, "Photonic band structure: observation of an energy gap for light in 3-D periodic dielectric structures", *Proc. Conference on Quantum Electronics and Laser Science* (*QELS'89*), Baltimore 1989 (OSA, Washington, D.C., 1989) TUKK6.
27) E. Yablonovitch, T. J. Gmitter, and K. M. Leung, "Photonic band structure: The face-centered-cubic case employing non-spherical atmos", submitted to *Phys. Rev. Lett.*
28) P. W. Anderson, "Ordering and Antiferromagnetism in Ferrites", *Phys. Rev.*, Vol. 102 (1956) pp. 1008–1013.

Chapter 6

Experiments on FM Noise Reduction

FM noise in each type of lasers can be reduced by the methods discussed in Chapter 5. The method of negative electrical feedback has been generally employed for most types of lasers. Since the lowest magnitude of FM noise has been realized in gas lasers and a method analogous to that for gas lasers has been used for dye lasers, experiments on these two types of lasers are described in 6.1. Among solid state lasers, the method pumped by semiconductor lasers exhibits a low FM noise. This method is also analogous to those for gas and dye lasers and will be described in 6.2. The method for semiconductor lasers is slightly different from those for other lasers. Several specific methods, e.g., optical feedback, have also been employed and will be described in 6.3.

6.1 Gas and Dye Lasers

Figures 6-1 (a) and (b) shows the experimental setup employed to achieve the highest frequency stability, in which He-Ne lasers of 633 nm wavelength were used [1]. Although the saturated absorption spectral line of I_2 vapor had ever been used as a frequency reference, it was replaced by a high-finesse Fabry-Perot interferometer. The reasons behind this change are: (1) A high *S/N* value of frequency demodulation can be obtained because the Fabry-Perot interferometer is a linear and passive optical element. (2) The resonance linewidth of a high-finesse Fabry-Perot interferometer can be narrower than 1 kHz, which is even narrower than those of spectral lines in some atomic or molecular resonances. (3) The low loss Fabry-Perot interferometer shows a very high efficiency (i.e. the ratio between input power and transmitted or reflected power), as high as 25%, even though the finesse is higher than 10,000.

In this setup, a high finesse Fabry-Perot interferometer was fabricated by using a Zerodur as a spacer (see 5.2) and was shielded by installing it into a vacuum chamber. E/O modulators were used as high-speed phase modulators in order to get the derivative of the resonance curve of the Fabry-Perot interferometer. The control bandwidth was expanded as wide as several MHz. The laser frequency was locked to the center frequency of the derivative curve, by which the square root of the Allan

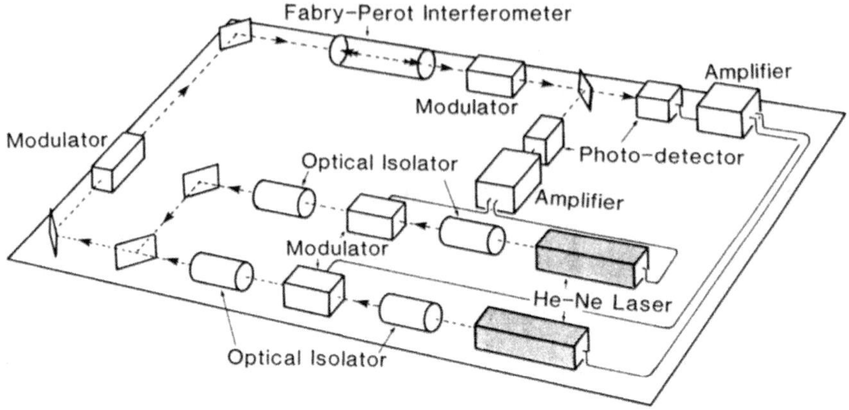

(a)

(b)

Fig. 6.1. Experimental setup for frequency control of a He-Ne laser. (a) Schematic explanation (after Salomon, Hils, and Hall [1]). (b) Photograph of the experimental setup of (a) (by the courtesy of J. L. Hall, NIST, 1991). A shield box hung on the ceiling is called "a quiet house".

variance for the residual FM noise was reduced to as low as 8×10^{-17} at the integration time of $\tau = 100$ s [2]. The result is shown in Fig. 6-2. This figure also shows that the phase fluctuations (PM noise) of the laser were reduced to less than 1 radian at $\tau < 2$ s, which means that a phase control was also achieved. Furthermore, as is shown in Fig. 6-3, the field spectral linewidth corresponding to the result in Fig. 6-2 was reduced to as low as 0.1 Hz [1]. However, this value has not yet reached the Schawlow-Townes limit of laser device or the shot noise limit of the control loop. It was estimated that the shot noise limit corresponded to a field spectral linewidth of 33 μHz. To obtain this value, however, drift in the reference angular frequency ν_r induced by the thermal expansion of Zerodur could give experimental difficulty. By carefully shielding the Fabry-Perot interferometer, the drift was reduced to as low as 170 kHz/day [2].

An Ar^+ laser has been used as the light source in a gravitational wave detection system, in which frequency stabilization was carried out, thus resulting in a power spectral density of 1×10^{-4} (Hz^2/Hz) at a Fourier frequency of 1 kHz [3]. However, a frequency stability higher than this value has not yet been reported owing to the large noise level induced by plasma instability in the discharge tube. Frequency stabilization in a waveguide-type CO_2 laser has been carried out by using a saturated absorption line in an intracavity NH_3 vapor cell as the frequency reference; as a result, a frequency stability of 1×10^{-11} was obtained at the integration time of 250 s [4]. This stabilized laser frequency was steadily swept within a range of 340 MHz by sweeping the voltage applied to the Stark electrodes in the NH_3 cell. A high finesse

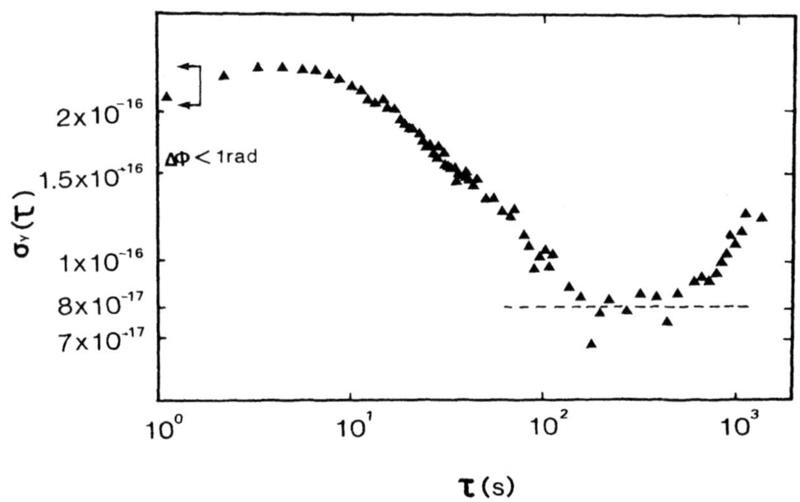

Fig. 6.2. Value of the square root of the Allan variance representing the stability of the center frequency of the field spectral shape of a He-Ne laser (after Hils and Hall [2]). For the integration time shorter than 2 s, the phase fluctuation was maintained less than 1 radian.

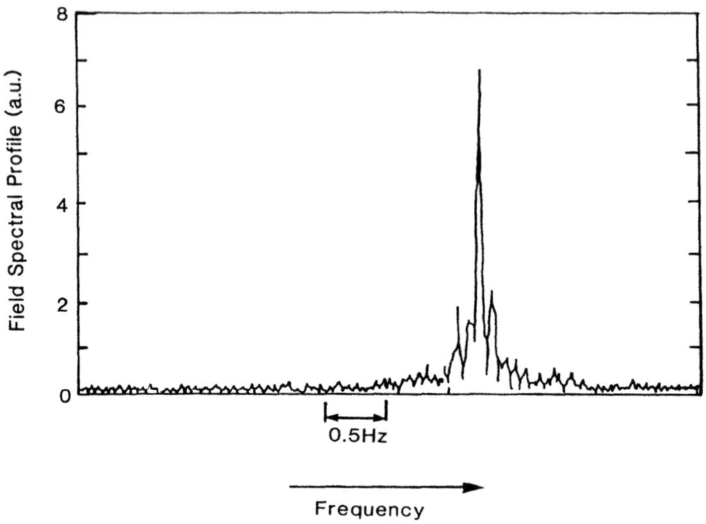

Fig. 6.3. Field spectral profile of a frequency-controlled He-Ne laser at 633 nm (after Salomon, Hils, and Hall [1]). FWHM of the spectral profile is about 0.1 Hz.

Fabry-Perot interferometer was used for the frequency stabilization of a dye laser, by which a field spectral linewidth narrower than 1 Hz was obtained [5].

6.2 Solid State Lasers

It has been experimentally confirmed that the magnitude of FM noise in a semiconductor laser-pumped YAG laser was far lower than that in a conventional YAG laser. Frequency stabilization of a semiconductor laser-pumped YAG laser was carried out by using an external high finesse Fabry-Perot interferometer and E/O modulator. Figure 6-4 shows the square root of the power spectral density of the FM noise [6]. Although curve B shows that the FM noise was reduced to a level lower than the shot noise limit of the frequency control loop (the curve C), this curve does not represent quantitatively the real magnitude of FM noise because it was estimated by measuring the magnitude of error signal fluctuations, i.e. fluctuations of the output signal from the frequency demodulator. However, this result means that the feedback loop gain employed in this system was very high.

The NPRO of Fig. 3-3(a) was very stable even in a free-running condition. Its field spectral was narrower than 10 kHz, and its frequency can be modulated by applying stress to the NPRO crystal by fixing PZT on the crystal; as a result, the modulation efficiency was about 450 kHz/V with a bandwidth of 500 kHz. A Fabry-Perot interferometer with a finesse larger than 20,000 was used for frequency control. The field spectral linewidth was reduced to 3 Hz and the square root of an

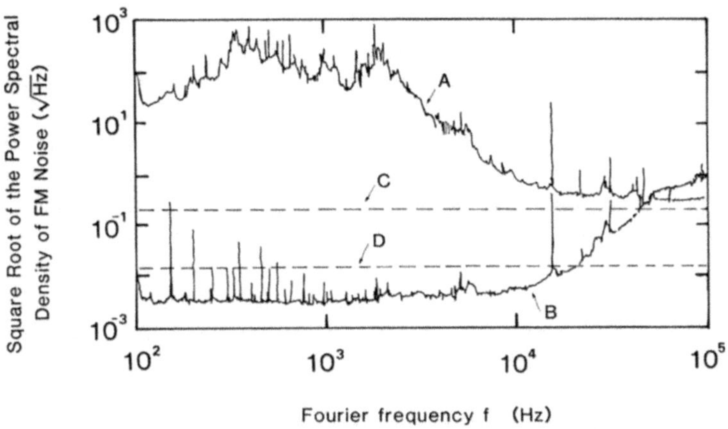

Fig. 6.4. Values of the square root of the power spectral density of the FM noise of a YAG laser pumped by a semiconductor laser (after Shoemaker, Brillet, Nary Man, Cregut, and Kerr [6]). The measured values were evaluated by the error signals from the control system. A: Free-running. B: With control. C: Schawlow-Towns' limit of the quantum noise of a laser determined by fluctuation of the spontaneous emission (corresponding to Eq.(4-49) and Π_s in Eq.(5-2)). D: Shot noise of the control system (corresponding to Π_n in Eq.(5-2)).

Allan variance of 1.5×10^{-14} was obtained at the integration time of 0.2 s [7]. Although these values reached the shot noise limit of the frequency control loop within a factor of 2, they have not yet reached to the Schawlow-Townes' limit of the free-running laser. However, these experimental results indicate that this type of laser is very stable and the compact and all solid-state laser is useful for a variety of application systems.

Although the magnitude of the FM noises of the gas, dye and solid-state lasers have been reduced to very small values, they have not yet reached the value of Π_s in Eq.(5-2), i.e. the coherent state. This means that an hyper-coherent state has not yet been achieved by these lasers. This is because the value of Π_s is smaller than Π_n for these lasers.

6.3 Semiconductor Lasers [8]

The FM bandwidth of a semiconductor laser is as wide as several GHz (see Fig. 3-30), which is much wider than that of other lasers. This fact allows us to design a high gain and wide bandwidth frequency control loop for semiconductor lasers. Since the five methods described in 5.1. have been employed in experiments, their results are reviewed in the following sections.

6.3.1 Negative electrical feedback

A very stable semiconductor laser system can be designed by employing the negative electrical feedback method. Figure 6-5 shows the block diagram of one of the systems of coherent light source, i.e., a synthesized system to achieve an optical sweep generator with low FM noise by using negative electrical feedback [9]. Although this system may look very complicated, it is more compact than those for other lasers and it can be expected that several electrical and optical components in this system will be integrated with semiconductor lasers on a common substrate. As is shown in Fig. 6-6, the five subjects described below can be simultaneously carried out in this system.

(a) Stabilization of the center frequency of the field spectrum

The drift of the center frequency in the field spectrum is reduced by reducing the FM noise at the Fourier frequency range lower than about 1 Hz by employing a DC servo-control loop. Several organic molecular vapors can be used as frequency references owing to the great number of absorption lines due to the higher harmonics and combination tones of vibration-rotation transitions in NH_3 [10], H_2O [11], C_2H_2 [12] and HCN [13], in the wavelength region of 0.8 ~ 1.6 μm. The problems associated with these vapors are that absorption coefficients are low, the Doppler-free narrow linewidth absorption spectra cannot be measured and spectral assignments are not easy. To solve these problems, the strong resonance lines of the Rb [14] and Cs [15] vapors due to electronic transitions around the 0.8 μm wavelength region have been used. They have several advantages, i.e. the volume of the experimental setup can be reduced because of their strong absorption, the

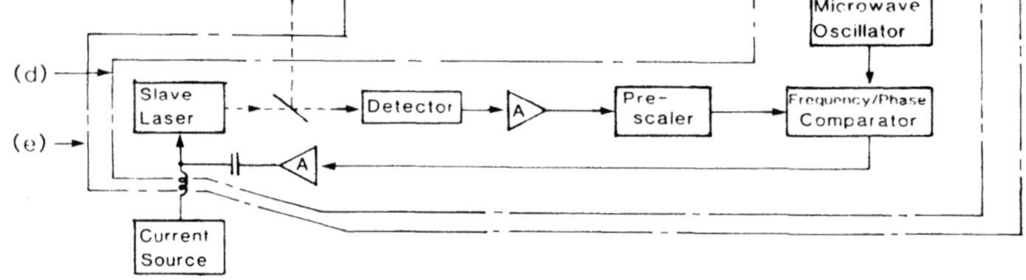

Fig. 6.5. A synthesized system to realize a low FM noise optical sweep generator by negative electrical feedback (after Ohtsu [9], ©1988 IEEE). Blocks (a) ~ (e) are for the stabilization of the center frequency of field spectrum, the accuracy improvement of the central frequency of field spectrum, the linewidth reduction of field spectrum, the frequency tracking, and the frequency sweeping, respectively.

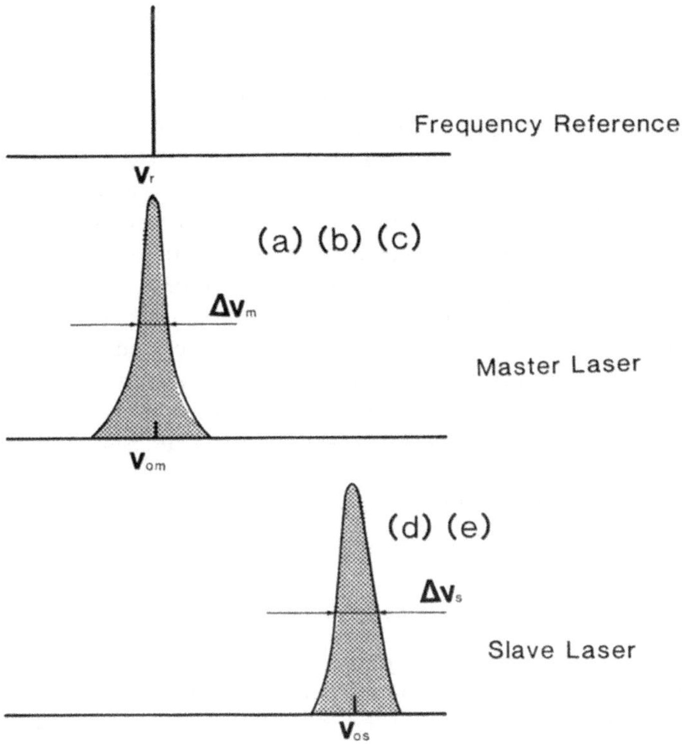

Fig. 6.6. Schematic explanation of (a) ~ (e) in Fig. 6.5. The frequency reference, and the spectral profiles of the master and slave lasers are shown. ν_r is the reference angular frequency. ν_{om} and ν_{os} are the center angular frequencies of the field spectral shapes of the master and slave lasers, respectively. $\Delta\nu_m$, $\Delta\nu_s$ are the linewidths (FWHM) of the master and slave lasers, respectively. (a) Stabilization of ν_m to ν_r. (b) Improvement of the accuracy of the stabilized angular frequency ν_m in (a). (c) Reduction of $\Delta\nu_m$. (d) Frequency tracking of the master laser to the slave laser. (e) Sweep of the ν_s with a high precision.

Doppler-free saturated absorption spectral lines can be measured and their spectral assignments have been completed. The signal strength of opto-galvanic spectral lines of several inert gases are also large [16]. Although a very high frequency stability cannot be realized due to the frequency shift in plasma instability of the discharge, a simple and practical frequency reference for 1.5 μm InGaAsP lasers has been proposed by using a miniature discharge lamp [17].

The InGaAsP laser has been used as a key device for optical communication systems and an advanced method for its frequency stabilization has been carried out by locking the second harmonic frequency generated from its active waveguide to the strong absorption spectral line in the Rb vapor [18]. Although the second harmonic power was as low as several pW, the sensitivity of this spectral measurement can be improved by using a heterodyning with a 0.78 μm AlGaAs laser. Furthermore, by using the AlGaAs laser as a pumping source for the Rb vapor, Doppler-free optical double resonance spectral lines in Rb can be measured by the second harmonics and they can be used as frequency references [19]. It can be claimed from these results that the low power of second harmonics does not give any essential problems in the frequency control of InGaAsP lasers. To generate a higher second harmonic power, nonlinear optical waveguides using organic materials have been fabricated [19]. For a 0.67 μm visible AlGaInP laser, resonance spectral lines in several atomic vapors can be used as frequency references, for example, the opto-galvanic spectroscopy of Li vapors has been carried out [20].

By comparing several properties of atomic and molecular spectral lines, the lines due to electronic transitions in atomic vapors have been considered to be more advantageous frequency references than those due to vibration-rotation transitions in molecular vapors because of absorption strength, simple spectral assignments, and so on. By using these frequency references, the square root of an Allan variance of frequency stability as high as 2×10^{-12} has been obtained [21]. It has been estimated that the shot noise limit of the frequency control loop corresponds to the frequency stability of 1×10^{-15} [9].

(b) Improvement of the accuracy of stabilized center frequency in the field spectrum

It is indispensable for applications in coherent optical systems to improve the frequency accuracy, i.e. to calibrate the laser frequency stabilized by the methods in (a). The most fundamental factor limiting accuracy is the drift in reference frequency. For example, the absorption spectral frequency of H_2O vapor exhibits a temperature coefficient with a shift of 3 MHz/K, which is due to the change in vapor pressure [22]. That of Rb vapor is 1/10 times that of H_2O vapor, from which it can be concluded that Rb can be used as a more accurate frequency reference [23].

Absolute measurements of optical frequency are essential in evaluating accuracy, for which a harmonic mixer using metal-insulator-metal (MIM) junction or Josephson junction point contact diodes has been employed [24]. However, it is difficult to use them for semiconductor lasers with a wavelength shorter than 1.5 μm because the sensitivity of the former and the response speed of the latter are low. Due to this difficulty, accuracy has been evaluated by conveniently measuring the optical wavelength instead of measuring the frequency. The accuracy evaluation has been carried out within an error of $1 \times 10^{-7} \sim 1 \times 10^{-8}$ by using a wavemeter composed of a scanning Michelson interferometer [10]. Several proposals for measuring optical frequency have been presented recently by fabricating a sensitive and fast photo-detector using a tunnel junction between the oxide superconductor film and the metal thin film [25] or using a fast response in virtual polarization in the quantum well semiconductor material [26]. If these devices can be used as a harmonic mixer, it is expected that the measurement error of the absolute frequency of the laser will be reduced to about 1×10^{-10}

(c) Linewidth reduction of the field spectrum

The FWHM of the field spectrum can be reduced if the feedback bandwidth B is wider than the FWHM $\Delta\nu_{FR}$ of the free-running laser. The reflection mode of a Fabry-Perot interferometer (see curve B of Fig. 5-6) has been used to achieve a high-gain and wide-bandwidth control loop. Figure 6-7 shows the FM noise power spectral density of a 0.8 μm AlGaAs laser, obtained by using an optimally designed control loop to maximize the gain and bandwidth and for which a network analysis was employed based on the transfer functions of servo-control loop elements [27, 28]. Curves A and B represent the magnitude of FM noise under free-running and frequency-controlled conditions, respectively. By comparing these curves, it is shown that the control bandwidth was 40 MHz and an FM noise reduction, as large as $1 \times 10^{-6} \sim 1 \times 10^{-7}$, was obtained within a Fourier frequency range of 100 Hz ~ 1 kHz. Curve C represents the FM noise level due to spontaneous emission fluctuations from the free-running laser i.e. the value of Eq.(4-49) which was estimated by assuming an α-parameter of 9. This level corresponds to the FM noise of a coherent state, or in other words, to Schawlow-Townes' limit. It can be found that the value of curve B is lower than that of curve C in a Fourier frequency lower than 4.4 MHz, i.e. a hyper-coherent state was achieved within this range for the first time by the negative electrical feedback of a semiconductor laser. The principal source of noise limiting the measurement of FM noise in this system was the IM

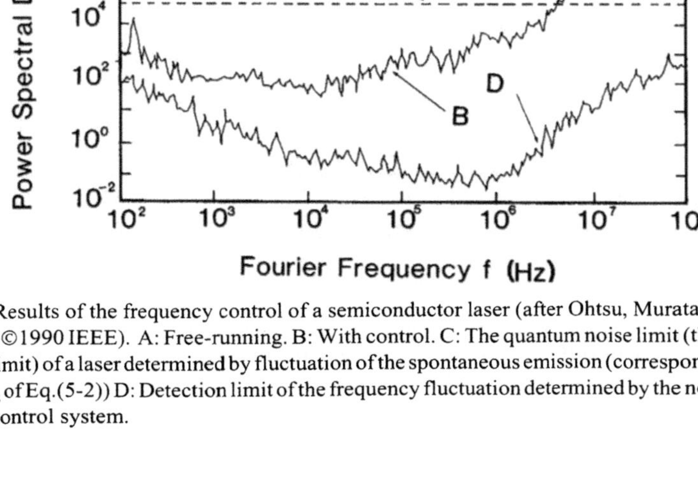

Fig. 6.7. Results of the frequency control of a semiconductor laser (after Ohtsu, Murata, and Kourogi [27], [28], ©1990 IEEE). A: Free-running. B: With control. C: The quantum noise limit (the Schawlow-Townes' limit) of a laser determined by fluctuation of the spontaneous emission (corresponding to Eq.(4-49) and Π_S of Eq.(5-2)) D: Detection limit of the frequency fluctuation determined by the noise generated from the control system.

noise of the laser, which is shown by curve D and corresponds to the Π_n of Eq.(5-2). It means that further improvement of the present feedback control system can reduce the FM noise to the level of the curve D.

Figure 6-8 shows a field spectral profile estimated by substituting the values

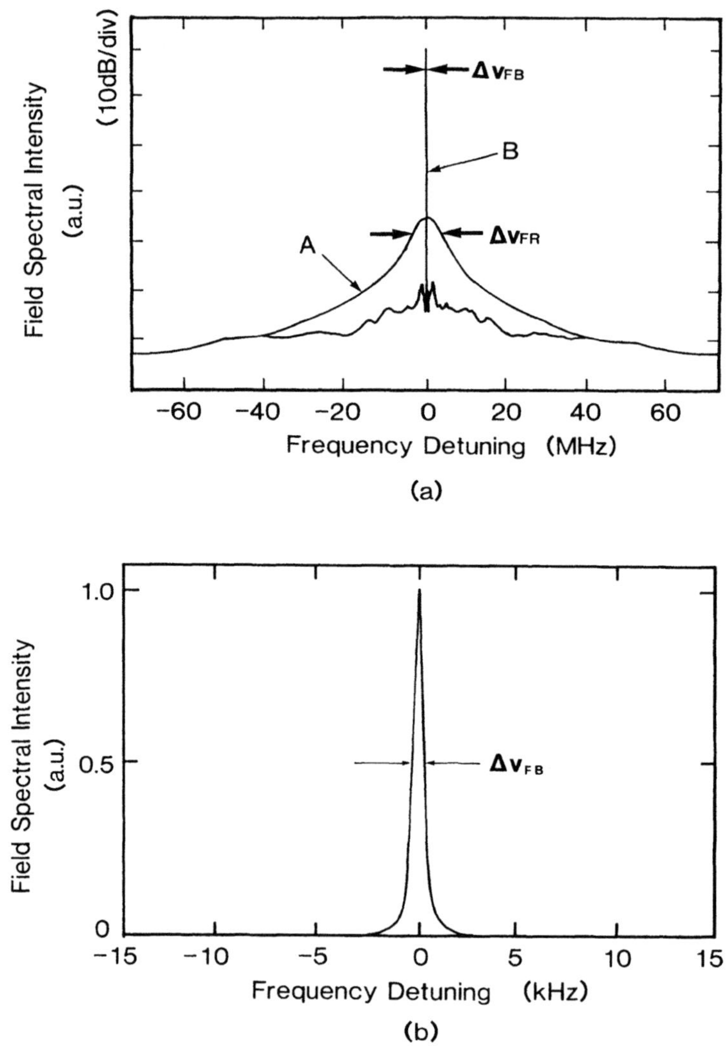

Fig. 6.8. Field spectral profile obtained from the measured results in Fig. 6.7 (after Ohtsu, Murata, and Kourogi [27], [28], ©1990 IEEE). (a) A: Free-running (corresponding to the curve A in Fig. 6.7). FWHM of the spectrum is 4.5 MHz. B: With control (corresponding to the curve B in Fig. 6.7). (b) Magnification of the curve B in (a). FWHM of the spectrum is 560 Hz.

of Fig. 6-7 into Eqs.(4-15) ~ (4-18). Curves A and B correspond to those in Fig. 6-7, respectively. These curves show that FWHM's are 4.5 MHz and 560 Hz for the free-running and frequency-controlled condition, respectively [27, 28]. A comparison of these two curves shows that the FWHM was reduced to 1/8,000 times that of the free-running laser. A FWHM of 1 Hz is expected if the FM noise is reduced to the value of curve D of Fig. 6-6. Furthermore, it has been estimated that the FWHM can be reduced to 50 mHz if the shot noise limit has been reached by removing the effect of laser power fluctuations, by using the frequency demodulator and the optical balanced detector, e.g. as shown in Fig. 5-8(a) [9].

Recently, the technique of optical feedback from an external Fabry-Perot cavity was employed as an auxiliary technique in negative electrical feedback system for further reduction of the linewidth. The experimental results are shown in Fig. 6-9 [29]. The power spectral density of residual FM noises was measured down to a Fourier frequency range of 50 mHz for the accurate evaluation of low FM noise components. Comparison between the curves A and B in Fig. 6-9(a) shows that the maximum FM noise reduction as high as 50 dB was achieved at the Fourier frequency of 10 kHz. By substituting the value of the curve B into Eqs.(4-15) ~ (4-18), the field spectral profile was estimated. As is shown in Fig. 6-9(c), it was confirmed that its FWHM was reduced to 7 Hz. The lines D in Figs. 6-9(a) and (b) represent the FM noise magnitude corresponding to the 1 Hz Lorentzian field spectral profile, which was estimated by using Eqs.(4-15) and (4-19). It is seen that the FM noise magnitude given by the curve B is lower than that of the line D for the Fourier frequency range of 10 Hz $\leq f \leq$ 1 MHz. It means that the field spectral profile can be regarded to be sub-Hz if this laser system is used for application systems which operate within the frequency range of 10 Hz $\leq f \leq$ 1 MHz.

(d) Frequency and phase tracking of a slave laser to a highly stable master laser

Frequency/phase tracking has been employed as an essential technique for heterodyne or homodyne optical detection system; this was realized by methods (a) ~ (c). Blocks (d) and (e) (Fig. 6-5) represent the frequency/phase locking loop, where the injection current of a slave laser is controlled by using an electronic frequency/phase comparator so that the frequency/phase of the beat signal between the two lasers is locked to that of the output signal from a microwave frequency synthesizer, i.e. a local oscillator [30]. This control can result in a heterodyne-type frequency/phase locked loop. As a result, the square root of the Allan variance of beat frequency fluctuations was reduced to as low as 1.7×10^{-15} at an integration time of 1000 s [30]. Based on this result, an improved system was constructed by using the optical feedback from the external Fabry-Perot cavity as an auxiliary technique (see Figs. 6-13 and 6-14) to obtain the results shown in Fig. 6-10. Figures 6-10 (a), (b) and (c) show the experimental results of the field spectral profile, waveform and square root of the Allan variance $\sigma_{y}(\tau)$ of the beat signal normalized to the optical frequency [31], which were obtained by following the design criterion of the phase locked loop [32]. The field spectral profile of Fig. 6-10(a) shows that

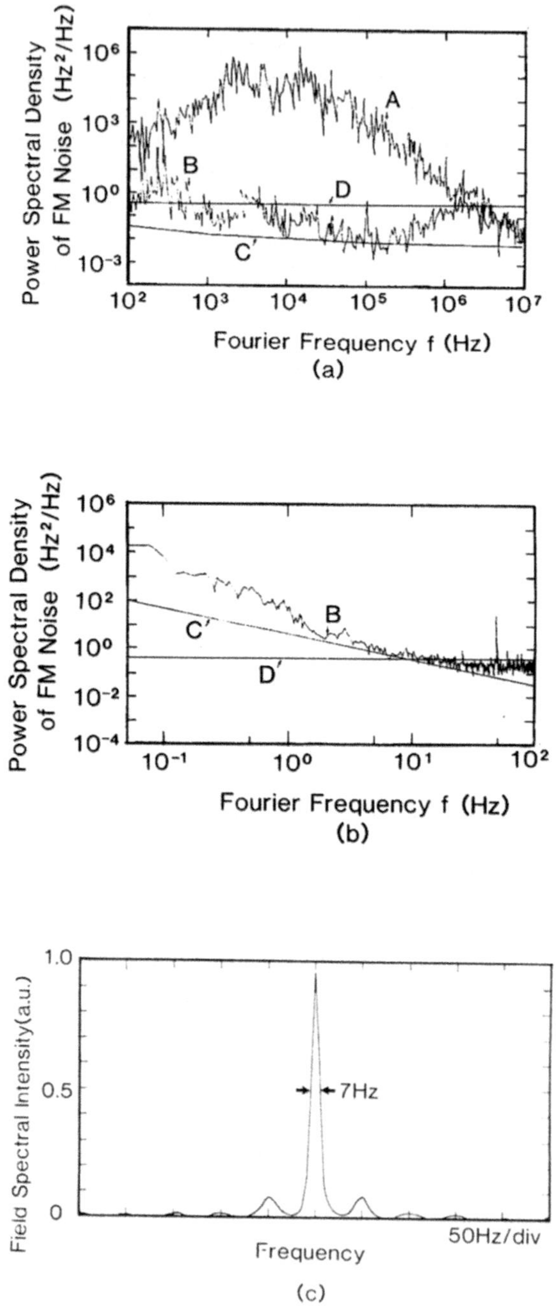
Power Spectral Density of FM Noise (Hz²/Hz)
10⁶
10³
10⁰
10⁻³
A
B
D
C′
10² 10³ 10⁴ 10⁵ 10⁶ 10⁷
Fourier Frequency f (Hz)
(a)
Power Spectral Density of FM Noise (Hz²/Hz)
10⁶
10⁴
10²
10⁰
10⁻²
10⁻⁴
B
C′
D′
10⁻¹ 10⁰ 10¹ 10²
Fourier Frequency f (Hz)
(b)
Field Spectral Intensity(a.u.)
1.0
0.5
0
7Hz
Frequency
50Hz/div
(c)

Fig. 6.9

the frequency/phase noise was drastically reduced up to the 1 MHz Fourier frequency range. The resolution of its FWHM measurement was limited by the resolution bandwidth (30 Hz) of the RF spectrum analyzer, from which it was estimated that the FWHM was far narrower than 30 Hz. Fig. 6-10(b) shows that the phase of the beat signal is locked tightly to that of the output signal from the microwave synthesizer. The measured value of $\sigma_y(\tau)$ of Fig. 6-10(c) (open circles) can be approximated as $6.3 \times 10^{-17}\tau^{-1}$ (the solid line). In particular, the value of $\sigma_y(\tau)$ was 1.1×10^{-18} at $\tau = 70$ s, which corresponds to 0.4 mHz beat frequency fluctuations. The broken curve corresponds to frequency fluctuations of the output signal from a microwave synthesizer. A comparison between this curve and the open circles shows that the stability of the beat signal is almost at the same level as that of the local oscillator, which implies a high tracking capability of the slave laser to the master laser. From this measurement, the phase error variance σ_ϕ^2 was estimated as 0.02 rad^2. For this estimation, the conversion formula $\sigma_\phi^2(\tau) = (\nu_0\tau)^2\sigma_y^2(\tau)$ was used, where ν_0 is the optical angular frequency. This value is smaller than 1 radian^2, which means that a heterodyne optical phase locking was obtained. In order to evaluate the performance of these frequency/phase locked loops, an Allan variance real-time processing system was developed [33] whose accuracy is 5000 times higher than that of previously developed systems [34]. A homodyne-type optical phase locked loop can also be achieved by fixing the local oscillator frequency to zero. In this case, the photo-detector used to mix the two lasers works as a phase comparator. Figure 6-11 shows the power spectral density of residual phase fluctuations in a homodyne-type optical phase locked loop [35]. By reducing the length of the control loop in the negative electrical feedback, the control bandwidth becomes as wide as 134 MHz. In this feedback loop, the loop length has been sufficiently shortened so that it is no longer the principal factor limiting the bandwidth. These factors are now the group delay of the laser device and the photo-detector, which means that a further expansion of the negative electrical feedback bandwidth can be expected by improving the performance of these devices. The phase error variance was estimated from this result as 0.15 rad^2.

(e) Stable and wideband frequency sweep

Accurate frequency sweep is achieved by utilizing an heterodyne-type frequency/phase locked loop as described in (d), i.e. the slave laser frequency is swept

Fig. 6.9. Result of the FM noise and linewidth reductions by adding the optical feedback from an external Fabry-Perot cavity to the negative electrical feedback as an auxiliary technique(after Shin and Ohtsu [29]). The curve C indicates the FM noise detection limit imposed by the IM noise of the laser. The line D represents the FM noise magnitude corresponding to the 1 Hz Lorentzian field spectral spectral profile, which was estimated by using Eqs.(4-15) and (4-19). (a) Power spectral density of the residual FM noise obtained by a lower gain and narrower bandwidth feedback loop (curve A), and a higher gain and wider bandwidth feedback loop (curve B). (b) Extension of curve B in (a) for the Fourier frequency range down to 50 mHz. (c) Calculated field spectral profile of the stabilized laser, where the curves B of (a) and (b) were used.

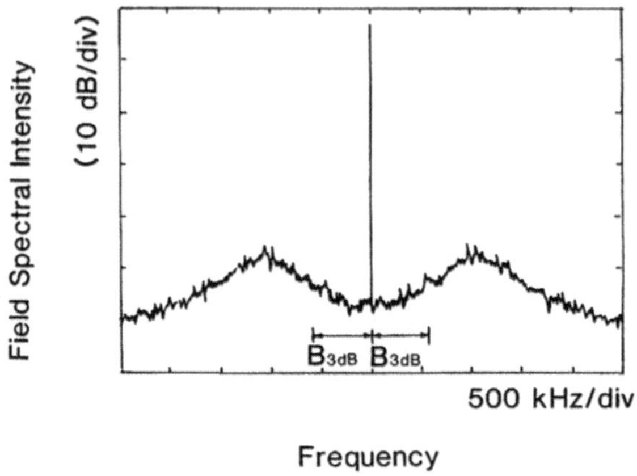
Field Spectral Intensity
(10 dB/div)
B3dB B3dB
500 kHz/div
Frequency

(a)

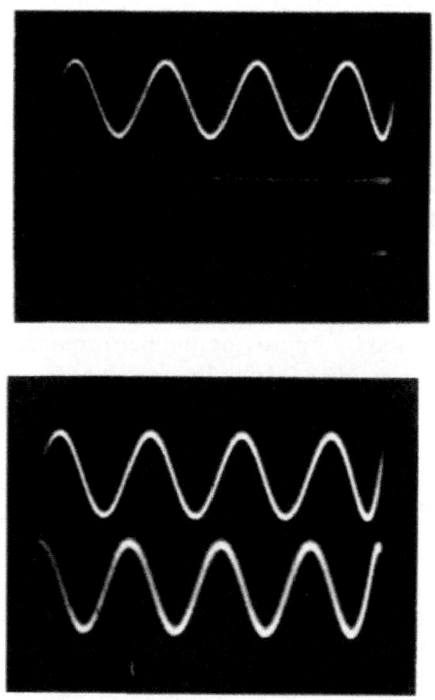

(b)

Fig. 6.10

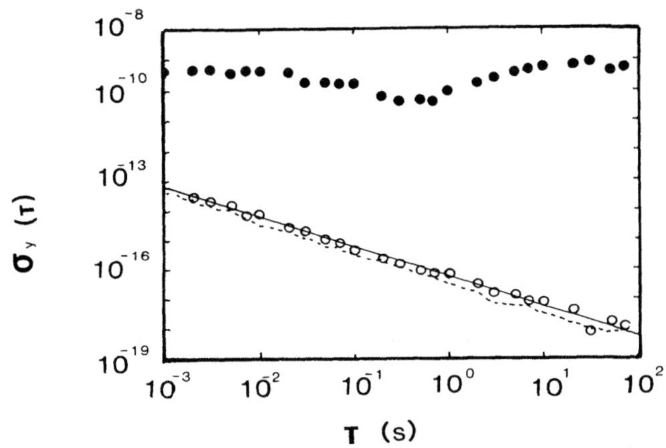

(c)

Fig. 6.10. Experimental results of the heterodyne phase locking (after Shin and Ohtsu [31]). (a) Field spectral shape of the heterodyne signal under phase locking condition. B_{3dB} is the calculated 3 dB bandwidth of the phase locked loop. The center frequency of the heterodyne signal was 25 MHz. (b) The heterodyne signals (lower traces) observed in the time domain by an RF oscilloscope under unlocking (upper picture) and phase-locking (lower picture) conditions. In these pictures, upper traces are the triggering signals from the reference signal to which the heterodyne signal was phase-locked. (c) Measured results of Allan variances of residual FM noise of the heterodyne signal. $\sigma_y(\tau)$ and τ are the square root of Allan variance and the integration time, respectively. The value of $\sigma_y(\tau)$ was normalized to the nominal optical frequency of the AlGaAs lasers used in this experiment. Closed circles, open circles, and a broken curve are the measured results for under free-running, under phase locking condition, and for the reference signal, respectively. A solid curve is a least-square approximation of the results for the phase-locked heterodyne signal, which is expressed as $\sigma_y(\tau) = 6.3 \times 10^{-17}\tau^{-1}$.

in a very stable manner by sweeping the microwave local frequency while maintaining a frequency/phase locked loop. A continuous frequency sweep range of 64 GHz, wide enough for high-resolution laser spectroscopy and optical length measurements, has been realized by using conventional Fabry-Perot cavity-type AlGaAs lasers [30]. A range as wide as 1 THz is achievable by using longitudinal mode-controlled lasers, e.g., DFB or DBR lasers, since this range is limited by the mode-hopping phenomenon of the laser devices used in this system [36]. It is also estimated that this range can be expanded to 700 THz by using the techniques of sum- or difference-frequency generation and parametric wavelength conversions by organic nonlinear waveguides, i.e. a Peta-Hz class optical sweep generator, are expected [37, 38]. A detailed explanation of this sweep generator is given in 8.1.2.

6.3.2 Reduction of the cavity loss

The first method to reduce cavity loss is to make use of optical feedback. In this case, the cavity loss Γ_s of Eq.(5-1) is reduced by modifying the cavity structure. The

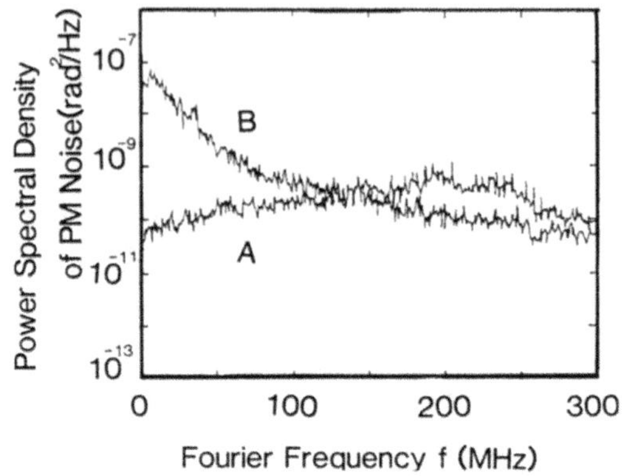

Fig. 6.11. Experimental results of the power spectral density of the residual PM noise of the homodyne phase locked loop obtained by using the negative electrical feedback to the two AlGaAs lasers (after Kourogi, Shin, and Ohtsu [35], ©1991 IEEE). A: Under wideband phase locked condition. B: Under the phase locked condition with about 10 MHz bandwidth.

integral on the right-hand side of the equation is not required to describe this feedback. A strong optical feedback is required and is obtained by coating an anti-reflection film on the laser facet to avoid deterministic instability in laser oscillation [39, 40], which has been called a coherent collapse [41]. A low-loss external cavity is formed by using a single external reflector to obtain a linewidth of about 2 kHz [42]. Although a wavelength sweep is achieved by using diffraction grating as an external reflector, a continuous sweep is difficult to obtain due to hopping between the longitudinal modes of the external cavity. In spite of these mode hoppings, a total wavelength sweep range of over 100 nm was demonstrated by fabricating a laser device with the introduction of a quantum well structure for this optical feedback [43].

The external cavity length should be controlled by a PZT in order to control the phase of reflected light and to achieve a stable optical feedback. The method of using a self-pumped phase conjugate mirror (a $BaTiO_3$ crystal) as an external reflector has been proposed to avoid this phase control [44]. (The principle of a phase conjugated mirror can be found in refs.[45, 46].) Although this type of optical feedback did not require any phase control owing to the self-aligning nature of the phase conjugate mirror, jumps between external cavity modes were observed; furthermore, the built-up time of the phase conjugated wave was too long for this application due to the slow response time of the $BaTiO_3$. To overcome this difficulty, it has been proposed to use a semiconductor laser device as a phase conjugate mirror [47]. As is shown in Figs. 6-12(a) and (b), since its reflectivity and bandwidth were as large as 10^3 and 1.9 GHz, respectively, a stable optical feedback can be expected by using this mirror.

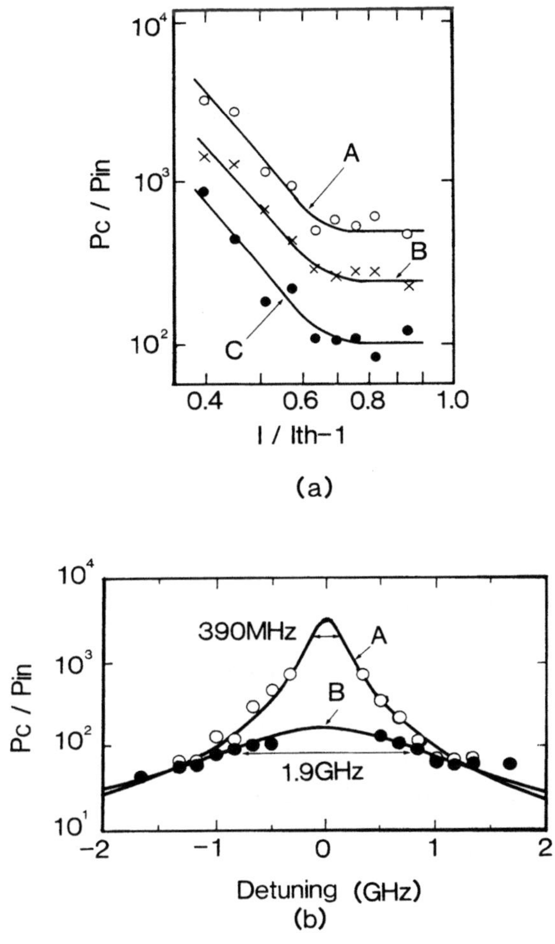

Fig. 6.12. Experimental results of the characteristics of an AlGaAs laser as a phase conjugate mirror (after Koshiishi, Ohtsu, and Teramachi [47]). (a) A relation between the reflectivity P_c/P_{in} of the phase conjugate mirror and the normalized bias level. The probe power P_{in} irradiated to the laser were −40.9 dBm (A), −36.2 dBm (B), and −30.9 dBm (C). P_c is the power of the phase conjugate wave. (b) Measured results of a relation between the reflectivity and the detuning between the pump and probe beams. Open and closed circles are for P_{in} = −32.2 and −25.0 dBm, respectively. The curves A and B are the Lorentzian, which were least-square-fitted to the experimental results.

As an alternative stable optical feedback, the method of using an external confocal Fabry-Perot cavity has been proposed [48]. An experimental setup is shown in Fig. 6-13, by which a linewidth of about 10 kHz was obtained. In this method, the transmitted light from the Fabry-Perot cavity after resonance was injected into the laser (the Fabry-Perot cavity worked as a frequency-dependent

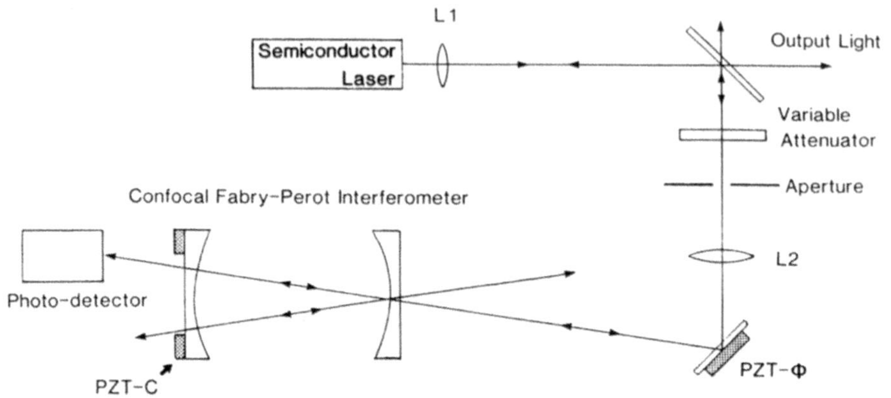

Fig. 6.13. Experimental setup for optical feedback from an external confocal Fabry-Perot cavity (after Dahmani, Hollberg, and Drullinger [48]).

reflector) and by which the laser frequency was pulled into the resonance frequency of the Fabry-Perot cavity resulted in a reduction in the FM noise of the laser. The bandwidth of this optical feedback is given by the photon lifetime of the Fabry-Perot cavity. The accuracy of controlling the separation between the laser and the Fabry-Perot cavity does not have to be as high as that in conventional optical feedback using a single external reflector [49]. In addition, this method has two outstanding advantages: (1) The solitary laser cavity structure is preserved since a strong optical feedback is not required; thus mode-hopping between any external cavity modes does not take place. (2) A frequency modulation is achievable by modulating the injection current with the frequencies related by rational fractions to the free-spectral range of the Fabry-Perot cavity [49, 50]. This algebraic relation corresponds to the Farey tree and to the devil's staircase [51], which can also be found in the mode locking of an external-cavity semiconductor laser [52].

Such a kind of Fabry-Perot cavity has been fabricated by employing the techniques of micro-optics and fiber-optics to realize a compact laser module as is shown by Fig. 6-14 [31, 53, 54]. This method of optical feedback was recently employed for a 0.67 μm wavelength visible InGaAlP laser to reduce the field spectral linewidth to less than 50 kHz [55], in which the linewidth under free-running condition had been as wide as 200 MHz.

The second method is to fabricate a new laser devices in order to improve lasing characteristics. Although lasing performances can be improved by adding several electrical or optical components to the negative electrical or optical feedback, system designers usually hesitate to add these external components since they prefer to keep the system structure as simple as possible. Some of the improved performances of these new laser devices could be low-to-medium control gain and medium-to-wide control bandwidth, for example. The latter is due to the shorter

Fig. 6.14. A Modulatable compact semiconductor laser module. A technique of micro-optics is used to fabricate the external Fabry-Perot cavity (by the courtesy of K. Nishide and T. Imai, Tokyo Aircraft Instrument Co. Ltd., 1989).

photon lifetime of the laser cavity as compared to those of the above-mentioned optical feedback systems. It is indispensable to use negative electrical feedback in these laser devices because their control gain is not high.

There can be two methods of developing new laser devices: (1) The cavity loss is decreased by increasing the length of the laser cavity. Furthermore, competitions among longitudinal modes are suppressed by reducing spatial hole-burning by controlling the magnitudes of currents injected into separate-contact electrodes. In addition to these, it is effective to reduce the value of the α-parameter by introducing a quantum well structure [56]. (2) Fabrication of an integrated version of negative electrical or optical feedback. The integration of an external reflector with a laser device has been demonstrated and corresponds to the optical feedback method [57]. Furthermore, to avoid deterministic instability in the case of using a single external reflector, the integration of the laser gain part, a DBR as external reflector and a phase modulator to control the phase of the reflected light have been tried and the spectral half linewidth has been reduced to as low as several 100 kHz [58, 59]. Two DBR's and an optical waveguide have been integrated, which correspondeds to the optical feedback in Fig.6-13. The linewidth of a 1.5 μm wavelength InGaAsP laser has been reduced to 150 kHz by this device [60]. Nowadays, the integration of these components with a laser could be the problem remaining to be solved. In the case of negative electrical feedback, the principal factor limiting the control bandwidth could be the feedback loop length which will be reduced by integrating the optical frequency demodulator, photo-detectors and amplifiers with the laser device. By this integration, the performance of the negative electrical feedback system can be improved, i.e. this system would then be compatible with opto-electronic integrated circuits.

6.3.3 Suppression and enhancement of spontaneous emission

FM noise can be reduced by reducing the value of Π_s on the right-hand side of Eq.(5-2), i.e. the magnitude of spontaneous emission fluctuations. For this purpose, the probability of spontaneous emission is controlled by confining the gain medium to a micro-cavity in which the number of resonant modes is limited. This field of study has been called cavity quantum electrodynamics (cavity QED) [61]. There are several problems remaining to be solved in this technique since it is still at preliminary stage of study. For example, spontaneous emission can be enhanced at the resonant frequency of the micro-cavity, which would be useful in developing a high quantum-efficiency light emitting device; however, in such a case, FM noise would be increased. On the other hand, spontaneous emission can be inhibited at a non-resonant frequency; however, the threshold of the laser oscillation would be increased.

Experiments have been carried out, mainly in the microwave frequency region, by using Rydberg atoms because the size of the micro-cavity does not have to be very small in this region. However, a very short Fabry-Perot cavity was also used recently [62]. Furthermore, a fabrication technique for micro semiconductor devices has made it possible to develop a micro-cavity in the optical frequency region [63]. As another approach, control of spontaneous emission by producing a band gap in the energy structure of the spatially localized photon which was confined in a medium whose refractive index is periodically modulated has been proposed [64, 65, 66]. It is expected that low noise and high quality lasers will be developed in the near future by utilizing these new phenomena of quantum optics.

[References]

1) Ch. Salomon, D. Hils, and J. L. Hall, "Laser stabilization at the millihertz level", *J. Opt. Soc. Am. B*, Vol. 5 (1988) pp. 1576–1587.

2) D. Hils and J. L. Hall, "Ultra-Stable Cavity-Stabilized Lasers with Subhertz Linewidth", *Frequency Standards and Metrology* (*Proceedings of the 4-th Symposium, Ancona, Italy, Sept. 5-9, 1988*) Ed. by A. De Marchi (Springer, Berlin, 1989) pp. 162–173.

3) G. A. Kerr, N. A. Robertson, J. Hough and C. N. Man, "The Fast Frequency Stabilization of an Argon Laser to an Optical Resonator Using an Extra-Cavity Electro-Optic Modulator", *Appl. Phys. B*, Vol. 37 (1985) pp. 11–16.

4) T. Nakamura, M. Ohtsu, and T. Tako, "Frequency Stabilization and Sweep for a Waveguide CO_2 Laser by Using Stark Absorption Lines", *J. Jpn. Soc. of Spectroscopy*, Vol. 34, pp. 23–28, 1985 (in Japanese).

5) J. L. Hall, M. Zhu, F. Shimizu and K. Shimizu, "External frequency stabilization of a commercial dye laser at the sub-hertz level", *Proc. of the 16th International Quantum Electronics Conference* (*IQEC'88*), Tokyo, 1988 (IEE Japan, Tokyo, 1988) TuP21.

6) D. Shoemaker, A. Brillet, C. N. Man, O. Cregut and G. Kerr, "Frequency-stabilized laser-diode-pumped Nd : YAG laser", *Opt. Lett.*, Vol. 14 (1989) pp. 609–611.
7) T. Day, E. K. Gustafson, and R. L. Byer, "Active frequency stabilization of a 1.062 μm, Nd : GGG, diode-laser pumped nonplanar ring oscillator to less than 3 Hz of relative linewidth", *Opt. Lett.*, Vol. 15 (1990) pp. 221–223.
8) M. Ohtsu and K. Nakagawa, "Frequency Control of Semiconductor Lasers and Its Applications", *Oyo-Buturi*, Vol. 58, pp. 1428–1444 (1989) (in Japanese).
9) M. Ohtsu, "Realization of Ultrahigh Coherence in Semiconductor Lasers by Negative Electrical Feedback", *J. Lightwave Technol.*, Vol. 6 (1988) pp. 245–256.
10) M. Ohtsu, H. Kotani, and H. Tagawa, "Spectral Measurements of NH_3 and H_2O for Pollutant Gas Monitoring by 1.5 μm InGaAsP/InP Lasers", *Jpn. J. Appl. Phys.*, Vol. 22 (1983) pp. 1553–1557.
11) K. Fukuoka, M. Ohtsu, and T. Tako, "Accurate Wavelength Meaurements of the Absorption Lines in H_2O Vapor by a 0.8 μm AlGaAs Laser", *Jpn. J. Appl. Phys.*, Vol. 23 (1984) pp. L117–L120.
12) S. Kinugawa, H. Sasada, and K. Uehara, "Detection of C_2H_2 absorption lines with 1.5 μm DFB lasers", Extended Abstracts (The 49th Autumun Meeting, 1988); The Jpn. Soc. Appl. Phys., October, 1988, 6pQ12 (in Japanese).
13) H. Sasada, "1.5 μm DFB semiconductor laser spectroscopy of HCN", *J. Chem. Phys.*, Vol. 88 (1988) pp. 767–777.
14) M. Hashimoto and M. Ohtsu, "Laser Spectroscopy and Frequency Stabilization of a Semiconductor Laser for a ^{87}Rb Atomic Clock", *Trans. IEE Jpn.*, Vol. 108-C (1988) pp. 706–712 (in Japanese).
15) T. Yabuzaki, A. Ibaragi, H. Hori, M. Kitano, and T. Ogawa, "Frequency Locking of a GaAlAs Laser to a Doppler-Free Spectrum of the Cs-D_2 Line", *Jpn. J. Appl. Phys.*, Vol. 20 (1981) pp. L451–L454.
16) S. Yamaguchi and M. Suzuki, "Frequency stabilization of a diode laser by use of the optogalvanic effect", *Appl. Phys. Lett.*, Vol. 41 (1982) pp. 597–598.
17) Y. C. Chung, "Frequency-Locking of a 1.3 μm DFB Laser Using a Miniature Argon Glow Lamp", *IEEE Photonics Technol. Lett.*, Vol. 1 (1989) pp. 135–136.
18) M. Ohtsu and E. Ikegami, "Frequency stabilisation of 1.5 μm DFB laser using internal second harmonic generation and atomic ^{87}Rb line", *Electron. Lett.*, Vol. 25 (1989) pp. 22–23.
19) E. Ikegami, H. Kusuzawa, K. Nakagawa, and M. Ohtsu, "Monlinear Organic Waveguides for a LD-based Optical Sweep Generator I (Frequency Tracking)", Extended Abstracts (The 50th Autumn Meeting, 1989), The Jpn. Soc. Appl. Phys., Sept., 1989, 28pZP16 (in Japanese).
20) M. Ohtsu, H. Suzuki, K. Nemoto, and Y. Teramachi, "Narrow-Linewidth Tunable Visible InGaAlP Laser, Application to Spectral Measurements of Lithium, and Power Amplification", *Jpn. J. Appl. Phys.*, Vol. 29, (1990) pp. L1463–L1465.
21) M. Ohtsu, "Frequency stabilization in semiconductor lasers", *Opt. Quantum Electron.*, Vol. 20, pp. 283–300, 1988.
22) V. Pevtshin and S. Ezekiel, "Investigation of Absolute Stability of Water-Vapor-Stabilized Semiconductor Laser", *Opt. Lett.*, Vol. 12 (1987) 172–174.
23) H. Furuta and M. Ohtsu, "Evaluation of frequency shift and stability in rubidium vapor stabilized semiconductor lasers", *Appl. Opt.*, Vol. 28 (1989) pp. 3737–3743.
24) D. A. Jennings, K. M. Evenson, and D. J. E. Knight, "Optical Frequency Measurement", *Proc. IEEE*, Vol. 74 (1986) pp. 168–178.
25) Y. Enomoto and T. Murakami, "Optical Detector Using Oxide Superconductor", Extended Abstracts (The 36th Spring Meeting, 1989); The Jpn. Soc. Appl. Phys. and Related Soc., April, 1989, 2pZD4 (in Japanese).
26) M. Yamanishi, "Ultrafast Optical Process through Virtual Charge Polarization in Quantum Well Structures", *Oyo-Buturi*, Vol. 58 (1989) pp. 1696–1706 (in Japanese).
27) M. Ohtsu, M. Murata, and M. Kourogi, "Subkilohertz linewidth of a semiconductor laser by electrical feedback and its network analysis", Proc. Conference on Lasers and Electro-Optics (CLEO'89), Baltimore 1989 (IEEE/LEOS and OSA, Washington, D.C., 1989) THK30.

28) M. Ohtsu, M. Murata, and M. Kourogi, "FM Noise Reduction and Subkilohertz Linewidth of an AlGaAs Laser by Negative Electrical Feedback", *IEEE J. Quantum Electron.*, Vol. 26 (1990) pp. 231–241.
29) S. -H. Shin and M. Ohtsu, "Stable semiconductor laser with a 7-Hz linewidth by an optical-electrical double-feedback technique", *Opt. Lett.*, Vol. 15 (1990) pp. 1455–1457.
30) K. Kuboki and M. Ohtsu, "A Synthesized Method to Improve Coherence in Semiconductor Lasers by Electrical Feedback", *IEEE J. Quantum Electron.*, Vol. 25 (1989) pp. 2084–2090.
31) C. -H. Shin and M. Ohtsu, "Heterodyne Optical Phase-Locked Loop by Confocal Fabry-Perot Cavity Coupled AlGaAs Lasers", *IEEE Photonics Technol. Lett.*, Vol. 2 (1990) pp. 297–300.
32) C.-H. Shin, K. Kuboki, and M. Ohtsu, "Simulation for the Heterodyne Type Optical Phase-Locked Loop by Semiconductor Lasers", *Trans. IEE Jpn.*, Vol. 108-C (1988) pp. 678–684 (in Japanese).
33) C. -H. Shin, M. Teshima, and M. Ohtsu, "Heterodyne Optical Phase-Locked Loop by Semiconductor Lasers", Extended Abstracts (The 50th Autumn Meeting, 1989); The Jpn. Soc. Applied Phys., Sept. 1989, 29pZL4 (in Japanese).
34) K. Kuboki and M. Ohtsu, "An Allan Variance Real-Time Processing System for Frequency Stability Measurements of Semiconductor Lasers", *IEEE Trans. Instrum. and Meas.*, Vol. 39 (1990) pp. 637–641.
35) M. Kourogi, C. -H. Shin, and M. Ohtsu, "A 134 MHz Bandwidth Homodyne Optical Phase-Locked-Loops of Semiconductor Laser Diodes", *IEEE Photon. Technol. Lett.*, Vol. 3 (1991) pp. 270–272.
36) Y. Tohmori, K. Komori, S. Arai, and Y. Suematsu, "Low-threshold-current CW operation of 1.5 μm GaInAsP/InP bundle-integrated-guide distributed-Bragg-reflector (BIG-DBR) lasers", *Electron. Lett.*, Vol. 21 (1985) pp. 743–745.
37) M. Ohtsu, "Progress toward Highly Coherent Semiconductor Lasers", Technical Digest of the seventh International Conf. on Integrated Opt. and Opt. Fiber Commun., (IOOC'89), Kobe (IEICE Japan, Tokyo, 1989) 19A3-1.
38) H. Kusuzawa, E. Ikegami, K. Nakagawa, and M. Ohtsu, "Nonlinear Organic Waveguides for a LD-based Optical Sweep Generator II (Frequency Sweep)", Extended Abstracts (The 50th Autumn Meeting, 1989); The Jpn. Soc. Applied Phys., Sept. 1989, 28pZP17 (in Japanese).
39) R. Lang and K. Kobayashi, "External optical feedback effects on semiconductor laser properties", *IEEE J. Quantum Electron.*, Vol. QE-16, pp. 347–355, 1980.
40) M. Ohtsu, "Noises in Lasers", *Physics Monthly*, Vol. 6 (1985) pp. 297–303 (in Japanese).
41) D. Lenstra, B. H. Verbeek, and A. J. den Boef, "Coherence collapse in single-mode semiconductor lasers due to optical feedback", *IEEE J. Quantum Electron.*, Vol. QE-21 (1985) pp. 674–679.
42) N. A. Olsson and J. P. Van der Ziel, "Performance characteristics of 1.5 μm external cavity semiconductor lasers for coherent optical communication", *J. Lightwave Technol.*. Vol. LT-5 (1987) pp. 510–515.
43) D. Mehuys, M. Mittelstein, and A. Yariv, R. Sarafaty, and J. E. Unger, "Optimised Fabry-Perot (AlGa) As quantum-well lasers Tunable over 105 nm", *Electron. Lett.*, Vol. 25 (1989) pp. 143–145.
44) K. Vahala, K. Kyuma, A. Yariv, S. K. Kwong, M. Cronin-Golomb, and K. Y. Kau, "Narrow linewidth, single frequency semiconductor laser with a phase conjugate external cavity mirror", *Appl. Phys. Lett.*, Vol. 49 (1986) pp. 1563–1565.
45) R. A. Fisher (ed.), *Optical Phase Conjugation*, Academic Press, New York, 1983.
46) D. M. Pepper, "Applications of optical phase conjugation", Scientific American, 1986, January, pp. 56–65.
47) M. Ohtsu, I. Koshiisi, and Y. Teramachi, "A Semiconductor Laser as a Stable Phase Conjugate Mirror for Linewidth Reduction of Another Semiconductor Laser", *Jpn. J. Appl. Phys.*, Vol. 29 (1990) pp. L2060–L2062.
48) B. Dahmani, L. Hollberg, and R. Drullinger, "Frequqncy stabilization of semiconductor lasers by resonant optical feedback", *Opt. Lett.*, Vol. 12 (1987) pp. 876–878.

49) M. Ohtsu, "A Frequency Modulatable Narrow-Linewidth Semiconductor Laser", IEICE Technical Report, (IEICE Jpn., Tokyo, Dec. 1987) OQE87-135 (in Japanese).
50) L. Hollberg and M. Ohtsu, "Modulatable narrow-linewidth semiconductor lasers", *Appl. Phys. Lett.*, Vol. 35 (1988) pp. 944–946.
51) D. Baums, W. Elsasser, and E. O. Gobel, "Farey Tree and Devil's Staircase of a Modulated External-Cavity Semiconductor Laser", *Phys. Rev. Lett.*, Vol. 63 (1989) pp. 155–158.
52) H. Fujii, T. Shiragaki, M. Morimoto, T. Kobayashi, and T. Sueta, "Sub-harmonic mode locking of a semiconductor laser", Extended Abstracts (The 36th Spring Meeting, 1989); The Jpn. Soc. Appl. Phys. and Related Soc., 2pX1 (in Japanese).
53) J. Yoshida, T. Imai, K. Nishide, C. -H. Shin, M. Teshima, and M. Ohtsu, "Modulatable High Coherent Lasers for Fiber Optic Gyroscope", Reports on Topical Meeting of the Laser Soc. Jpn., (Laser Soc. Jpn., Osaka, 1989) Feb. 1989, RTM-88–37 (in Japanese).
54) T. Imai, J. Yoshida, K. Nishide, C. -H. Shin, M. Teshima, and M. Ohtsu, "High Coherent Semiconductor Laser for Fiber Gyroscope", Extended Abstracts (The 36th Spring Meeting, 1989); The Jpn. Soc. of Appl. Phys. and Related Soc., (The Jpn. Soc. Appl. Phys., Tokyo, 1989), March 1989, 4pZC2 (in Japanese).
55) M. Ohtsu, H. Suzuki, K. Nemoto, and Y. Teramachi, "Narrow-Linewidth Tunable Visible InGaAlP Laser, Application to Spectral Measurements of Lithium, and Power Amplification", *Jpn. J. Appl. Phys.*, Vol. 29 (1990) pp. L1463–L1465.
56) Y. Arakawa and A. Yariv, "Quantum Well Lasers - Gain, Spectra, Dynamics", *IEEE J. Quantum Electron.*, Vol. QE-22 (1986) pp. 1887–1899.
57) T. Fujita, J. Ohya, K. Matsuda, M. Ishino, H. Sato, and H. Serizawa, "Narrow spectral linewidth characteristics of monolithic integrated-passive-cavity InGaAsP/InP semiconductor lasers", *Electron. Lett.*, Vol. 21(1985) pp. 374–376.
58) I. Mito and K. Kitamura, "GaInAs/GaInAsP multiquantum well DFB lasers", Technical Digest, Conf. on Lasers and Electro-Optics (CLEO'89), Baltimore, 1989 (LEOS/IEEE, OSA, Washington, D.C., 1989) TUD5.
59) H. Imai, "Optical devices for high speed transmission including widely tunable DFB/DBR lasers", Technical Digest, Conf. on Lasers and Electro-Optics (CLEO'89), Baltimore, 1989 (LEOS/IEEE, OSA, Washington, D.C., 1989) FB4.
60) N. A. Olsson, C. H. Henry, R. F. Kazarinov, H. J. Lee, B. H. Johnson, and K. J. Orlowsky, "Narrow linewidth 1.5 μm semiconductor laser with a resonant optical reflector", *Appl. Phys. Lett.*, Vol. 51 (1987) pp. 1141–1142.
61) S. Haroche and D. Kleppner, "Cavity Quantum Electrodynamics", Phys. Today, January (1989) pp. 24–30.
62) A. Anderson, S. Haroche, E. A. Hinds, W. Jhe, D. Meschede, and L. Moi, "Atomic physics experiments in micron-sized cavities: suppressing spontaneous emission decay at optical frequency", Proc. of the 15th International Quantum Electronics Conference (IQEC'87), Baltimore, 1987 (LEOS/IEEE, OSA, Washington, D.C., 1987), THAA1.
63) J. L. Jewell, S. L. McCall, Y. H. Lee, A. Scherer, A. C. Gossard, and J. H. English, "Lasing characteristics of GaAs microresonators", *Appl. Phys. Lett.*, Vol. 54 (1989) pp. 1400–1402.
64) E. Yablonovitch, "Inhibited spontaneous emission in solid-state electronics", *Proc. of the 15th International Quantum Electronics Conference (IQEC'87)*, Baltimore, 1987 (LEOS/IEEE, OSA, Washington, D.C., 1987) THAA4.
65) E. Yablonovitch, "Photonic band structure: observation of an energy gap for light in 3-D periodic dielectric structures", *Proc. Conference on Quantum Electronics and Laser Science (QELS'89)* Baltimore, 1989 (LEOS/IEEE, OSA, Washington, D.C., 1989) TUKK6.
66) E. Yablonovitch, "Inhibited Spontaneous Emission in Solid State Physics and Electronics", *Phys. Rev. Lett.*, Vol. 58 (1987) pp. 2059–2062.

Chapter 7

Applications of Highly Coherent Lasers

There have been several application systems for which the use of highly coherent lasers is indispensable. Since the number of these systems is large, representative examples, especially those using recently improved semiconductor lasers, are reviewed in this chapter.

7.1 Optical Communication Systems

The development of optical communication systems has a long history. The proposal of using lasers in optical communication systems and basic experiments with those where suggested and carried out soon after the advent of lasers [1]. Moreover, it should be noted that the initial motivation force behind the invention of lasers was the requirement to increase the transmission capacity of communication systems. Especially after the development of glass fiber with a transmission loss as low as 20 dB/km and of room-temperature and CW semiconductor lasers in 1970, research and development in optical communication systems showed rapid progress. However, since the magnitude of FM noise in the semiconductor lasers was very large at that time, the intensity modulation and demodulation scheme had only been employed for signal transmission in practical systems. This scheme, therefore, did not satisfy the request to increase transmission capacity by utilizing the high optical frequency and high temporal coherence of lasers.

However, it was shown in 6.3 that low FM noise semiconductor lasers were recently developed and that the stable modulation of their frequencies or phases becomes possible, both of which could fully be utilized as advantageous properties of laser lightwaves. Furthermore, as is shown in Fig. 7-1, by using a heterodyne or homodyne demodulation scheme, the detection of sensitivity can become much higher than that in a conventional direct demodulation scheme, it also means that the signal transmission span can be increased. In this case, it is advantageous to use a 1.55 μm wavelength laser at which the optical fiber shows a minimum transmission loss, as low as 0.2 dB/km. The system employing these modulation and demodulation schemes has been called coherent optical communication. Practical systems

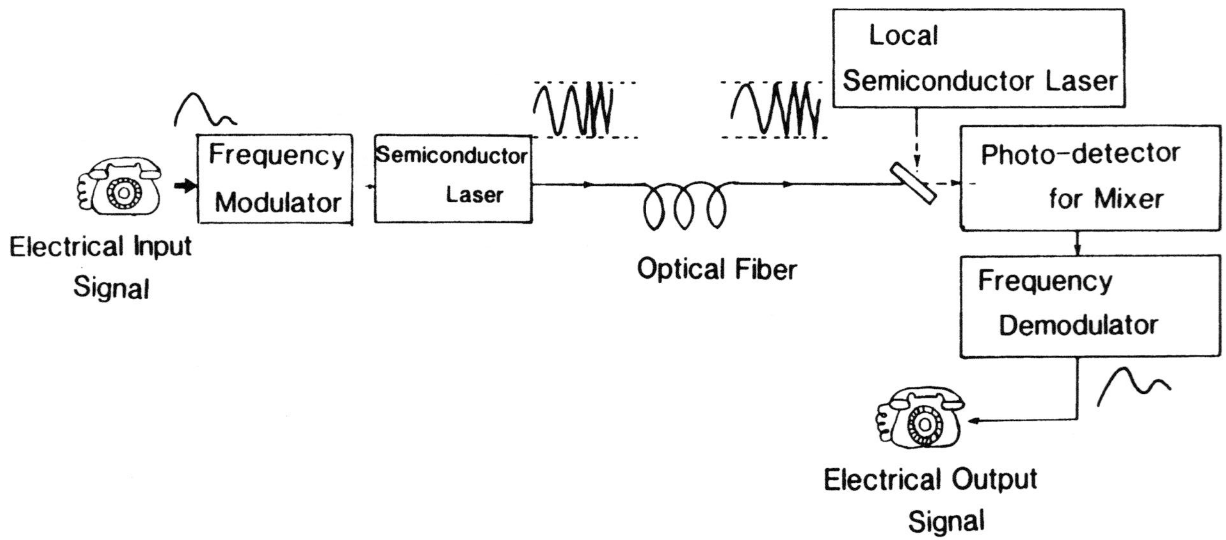

Fig. 7.1. Schematic configuration of a coherent heterodyne optical fiber communication system.

Fig. 7.2. The appearance of a coherent optical communication system (by courtesy of T. Nakagami, Fujitu Labs. Ltd., 1989).

have been recently realized by using a highly coherent semiconductor lasers. Figure 7-2 shows the appearance of such a coherent optical communication system.

A wider amplifier bandwidth is required for signal demodulation and is obtained by increasing the bit rate of the coherent optical transmission. Since it is rather difficult to design a wideband amplifier, it has been proposed to increase the transmission capacity by increasing the transmission channels instead of the bit rate. For these FDM (frequency division multiplexing) systems, channel numbers can be increased to as large as several thousands with several GHz channel-separation by using the 20THz wide low-loss window of the optical fiber around the 1.55 μm wavelength [2]. In this case, it is essential that the lasers satisfy the requirement imposed in 6.3.1(a), (b) and (e). In this FDM scheme, a large transmission capacity can be expected even though the transmission bit rate of each channel is not so high. Since (c) showed that a gain-bandwidth product as high as 40 THz was achieved for the reduction of FM noise in the feedback loop, it has been confirmed that the bit error rate can be kept sufficiently low when this laser is used for low-to-medium bit rate coherent transmission systems [3]. Furthermore, the requirement (d) should be satisfied for heterodyne or homodyne demodulation. Among them, the homodyne phase locked loop is the most important. Since it is still difficult to use an optical phase locked loop for practical systems owing to the insufficient temporal coherence in lasers, an alternative technique of phase diversity has been proposed to relax the requirement of temporal coherence in lasers [4].

Rapid progress in coherent optical communication systems in space has also occurred in intersattelite or satellite-to-ground communication. As they are similar to the coherent laser radar, 0.8 μm AlGaAs lasers have been used and controlled according to the techniques described in 6.3 [5]. The main problems remaining to be solved are increasing laser power and laser beam tracking.

7.2 Optical Measurements

(1) Passive ring resonator-type fiber gyroscope: A gyroscope is used in aircrafts for inertial navigation and to maintain the correct route towards the desired destination. Although conventional mechanical gyroscopes exhibit quite a high sensitivity, their inherent disadvantages are their short lifetime, sensitivity to mechanical shock and high price.

The development of optical gyroscopes goes rapidly, in order to overcome these difficulties, by using laser and optical fibers. Figure 7-3 shows the block diagram of a fiber gyroscope [6], which is operated based on the Sagnac effect [7]. That is, the phases of two output lights after traveling in clockwise and counter-clockwise directions in the fiber loop are different from each other if the fiber loop rotates around the axis perpendicular to the fiber loop plane. A rotation rate can be detected by measuring the phase difference between the two traveling lights. This fiber gyroscope has several advantages, e.g., its small size, light weight and long lifetime. The development of practical apparatus has advanced rapidly.

Figure 7-2 shows one of the most sensitive systems, possibly a promising candidate for the next generation of fiber gyroscopes. A ring Fabry-Perot cavity was formed by an optical fiber and directional couplers to realize clockwise and counterclockwise ring resonances. The Sagnac effect gives the difference in resonance frequencies between these two running lightwaves and can be sensitively measured if the finesse of the ring cavity is high. In this case, a semiconductor laser should be used to reduce the fiber loss and its field spectral linewidth should be narrower than 100 kHz in order to keep a high sensitivity [6]. The appearance of such a highly sensitive gyroscope is shown by Fig. 7-4. Experimental results have shown that a rotation rate lower than the revolution of the earth could be easily detected by reducing the FM noise of the laser [6, 8]. Furthermore, a compact and coherent semiconductor laser module has been fabricated for use in this system as is shown by Fig. 6-14 [9, 10, 11]

(2) Two-frequency doppler velocimeter: In a heterodyne laser interferometer, the value of the heterodyne frequency has limited the maximum measurable speed of translating objects. A typical system using a commercially available gas laser exhibits a maximum measurable speed of 1.8 m/s because the maximum heterodyne frequency is 20 MHz [12]. However, the heterodyne frequency can be increased up to 2 GHz by using the technique described in 6.3.1 for the master and slave lasers, which can increase the maximum measurable speed to 100 times obtained by conventional systems. This system could be effectively used to control

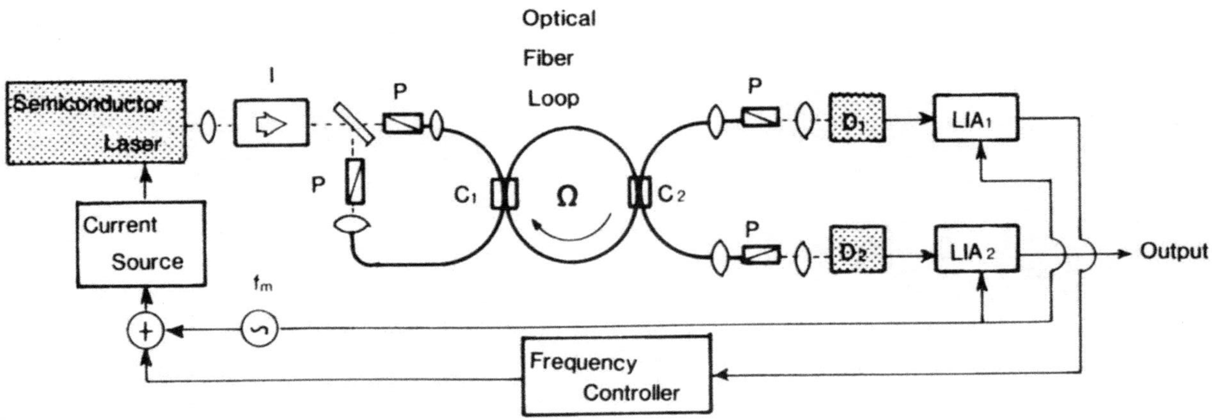

Fig. 7.3. Configuration of the ring-resonator type optical fiber gyroscope (after Ohtsu and Araki [6]). I: optical isolator. P: Polarizer. C_1, C_2: Fiber couplers. D_1, D_2: Optical detectors. LIA_1, LIA_2: Lock-in amplifiers.

Fig. 7.4. The appearance of a passive ring resonator-type fiber gyroscope system (by the courtesy of K. Nishide and T. Imai, Tokyo Aircraft Instrument Co. Ltd., 1989).

the speed of a fast stepper machine used in semiconductor factories if the X-ray exposure technique is employed in the future.

(3) Heterodyne displacement measurement: For this purpose, a high power and stable CO_2 laser has been conventionally used so far. The phase of the output lightwave was modulated by an external phase modulator to measure the displacement and velocity of a moving object. However, the system's performance was limited by the speed and efficiency of the phase modulator in the CO_2 laser. Since semiconductor lasers show a wideband and highly efficient direct FM characteristics in realizing versatile modulation schemes, it can be expected that the performance of this system will be improved.

(4) Photon scanning tunneling microscope: After the advent of ultrahigh resolution scanning tunneling microscopes (STM) utilizing the tunneling of electrons between the probe-tip and the conductive sample surface [13], a variety of STM has been developed such as atomic force STM [14]. The photon STM (PSTM) is one of the modified versions of STM and is a sort of super-resolution optical microscope used to measure the three-dimensional sample topography within the dimensions of an optical wavelength. This can be advantageously used to measure the profiles of a biomembrane, organic micro-thin film and nonconductive materials with non-contact and non-destructive manners. The PSTM can be considered as one of the versions of a near field microscopes (NFM) [15]; however, conventional NFM's have not shown a sufficiently high performance since basic technolo-

gies have not yet fully matured. Although more practical systems have been proposed recently [16], the system's performance is still at a preliminary stage of development. There can be two directions taken in the system's design:

(1) A transmission-type PSTM inducing a photo-chemical reaction in the sample in order to observe the result by high-resolution. This typical system is shown in Fig. 7-5(a) [17], and it is similar to the systems developed by Oak Ridge National Lab. [16]. In this figure, semiconductor lasers are used to carry out analytical spectroscopy at each local point on the sample.

(2) A reflection resonance-type PSTM improving resolution while maintaining non-contact and non-destructive conditions. This system is shown in Figs. 7-5(b) and (c) [17, 18].

A transmission-type PSTM picks up evanescent light from the sample surface by its aperture with a sub-wavelength diameter. Since the profile of the evanescent light field depends on sample topography, the three-dimensional profile of the sample can be measured. In Fig. 7-5(a), an optical fiber tip was used as aperture. Figure 7-6 shows an SEM picture of the Cr-film coated fiber tip, whose top was sharpened with a curvature radius of about 80 nm. The aperture is fabricated by removing the Cr-film from the top of the sharpened fiber tip. Since the picked up evanescent light can be interpreted as a tunneled photon, the position squeezing of the tunneled photon is achieved by a small aperture. This implies that the lateral resolution is determined by the diameter of the aperture. A high normal resolution is obtained since the probability of tunneling photons decreases exponentially with increasing sample-fiber separation. Figures 7-7(a) and (b) show example of measurement of a 9 nm thick SiO_2 film coated on half the prism's surface [17]. Although the picked up power was as low as 10 pW, a sensitive detection was carried out by directly modulating the semiconductor laser power and by employing phase sensitive detection. Even though the time constant of the phase sensitive detection was as short as 1 ms, a normal resolution of 2 nm was obtained. Figure 7-7(c) shows the measured result of spherical latex particles of 80 nm diameters, which are fixed on a prism's surface [17]. Even though the latex particles bunch due to their intrinsic chemical properties, the image of each particle can be clearly seen. However, it should be noted that the images are deformed and are oblates as a result of the bunching. From these results of measurements, the lateral resolution can be roughly estimated as 10 nm or even higher. Further improvements in resolution are expected by optimization of the modulation and demodulation parameters.

Because the transmission-type PSTM detects a very weak evanescent light power, its resolution can be limited by the detection sensitivity. To overcome this difficulty, the reflection resonance-type PSTM makes use of a short fiber Fabry-Perot cavity (Fig. 7-5(b)) or a semiconductor laser (Fig. 7-5(c)) as its pick up. It is also possible to replace the short fiber Fabry-Perot cavity by a low-loss dielectric optical waveguide grown on a glass substrate. This Fabry-Perot cavity is used as an external reflector to reduce the linewidth of the semiconductor laser by the optical feedback shown in Fig. 6-13. A small aperture is made on the pick-up surface of

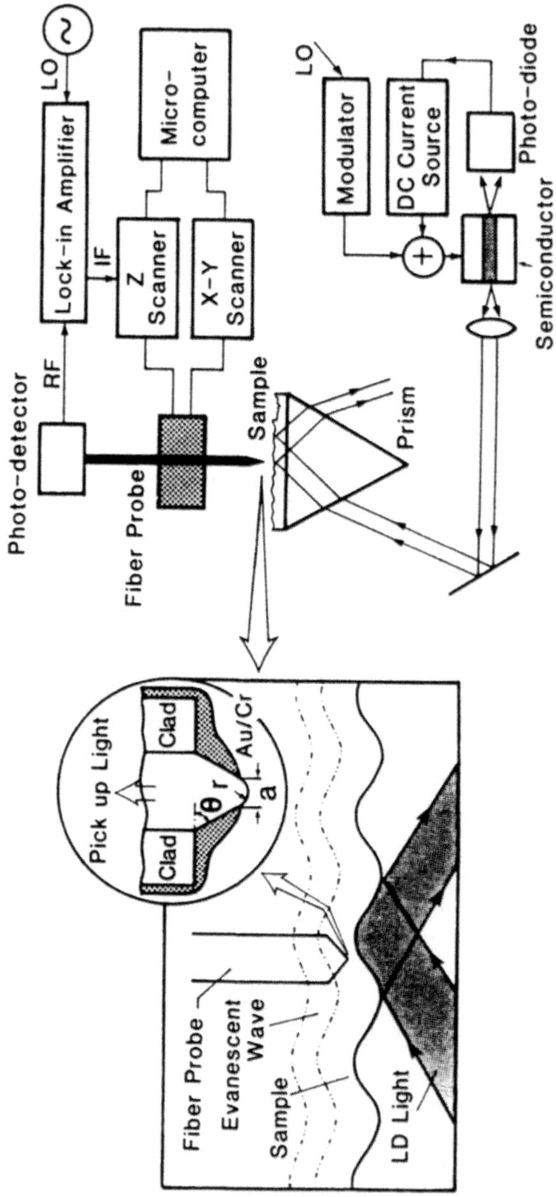
Photo-detector
RF
Lock-in Amplifier
LO
IF
Z Scanner
X-Y Scanner
Micro-computer
Fiber Probe
Sample
Prism
Modulator
DC Current Source
Photo-diode
Semiconductor Laser
Pick up Light
Clad
Clad
θr
a
Au/Cr
Fiber Probe
Evanescent Wave
Sample
LD Light

(a)

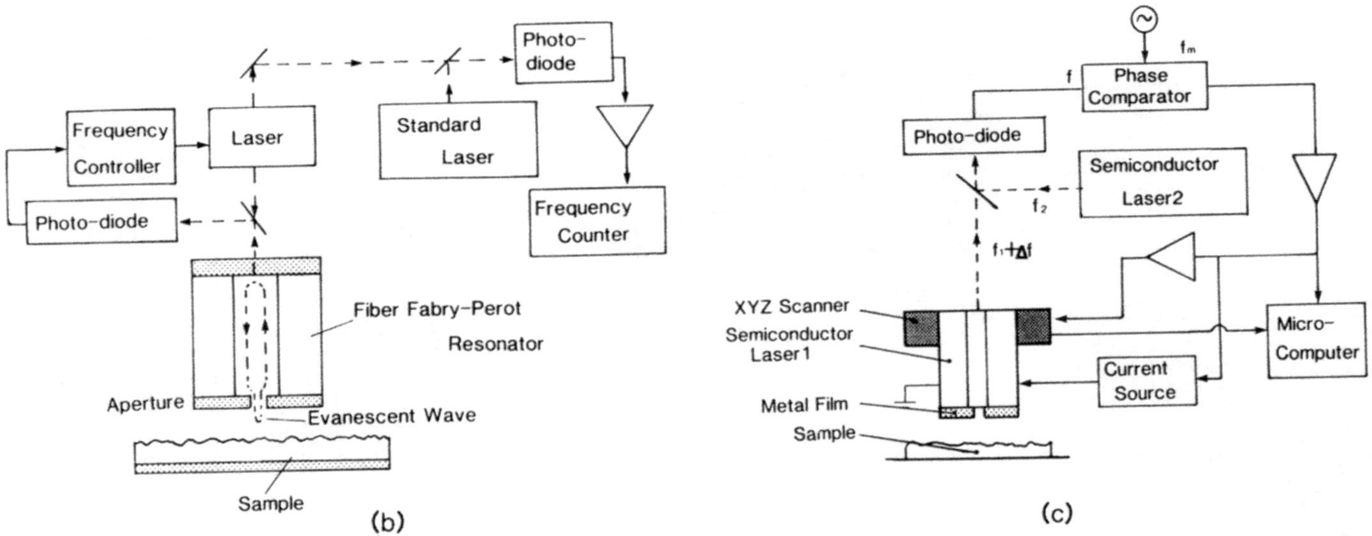

Fig. 7.5. Experimental setups of photon scanning tunneling microsocpe(after Ohtsu, Nakagawa, Jiang, and Tomita [17], and Jiang, Tomita, Nakagawa, and Ohtsu [18]). (a) Transmission-type. (b) Reflection-resonance-type using a fiber Fabry-Perot cavity as a pick-up. (c) Reflection-resonance-type using a semiconductor laser as a pick-up.

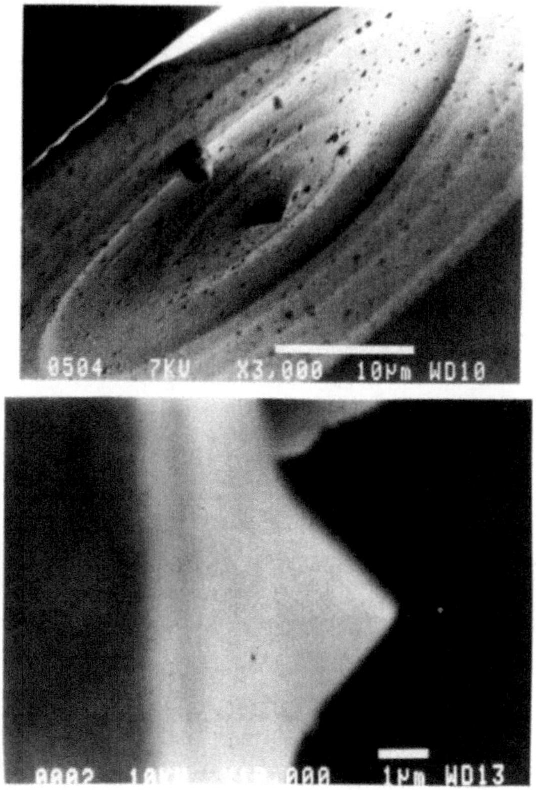

Fig. 7.6. A SEM pictures of a Cr-film coated fiber tip (after Ohtsu, Nakagawa, Jiang and Tomita [17]). Lower is the magnified picture of the upper.

Figs. 7-5(b) and (c). Since the resonance frequency of the cavity is slightly shifted when the evanescent light is reflected back from the sample, the three-dimensional profile can be measured by measuring this frequency shift. Since the technique of heterodyne optical phase locking can be employed to measure the frequency shift and since the residual heterodyne FM noise has been reduced to as low as 1×10^{-18}, quite a high resolution can be expected. The lateral resolution is determined by the diameter of the aperture, as was the case in the transmission-type PSTM. Figure 7-8 shows the calculated relationship between the aperture diameter and the normal resolution as a function of the residual heterodyne FM noise [17]. It can be seen from this figure that a normal resolution higher than 1 nm is expected, which is a drastic improvement of the resolution previously obtained. Actually, preliminary experiments have shown a normal resolution as high as $5 \times 10^{-3}\ \lambda$, which corresponds to several angstroms [17, 18].

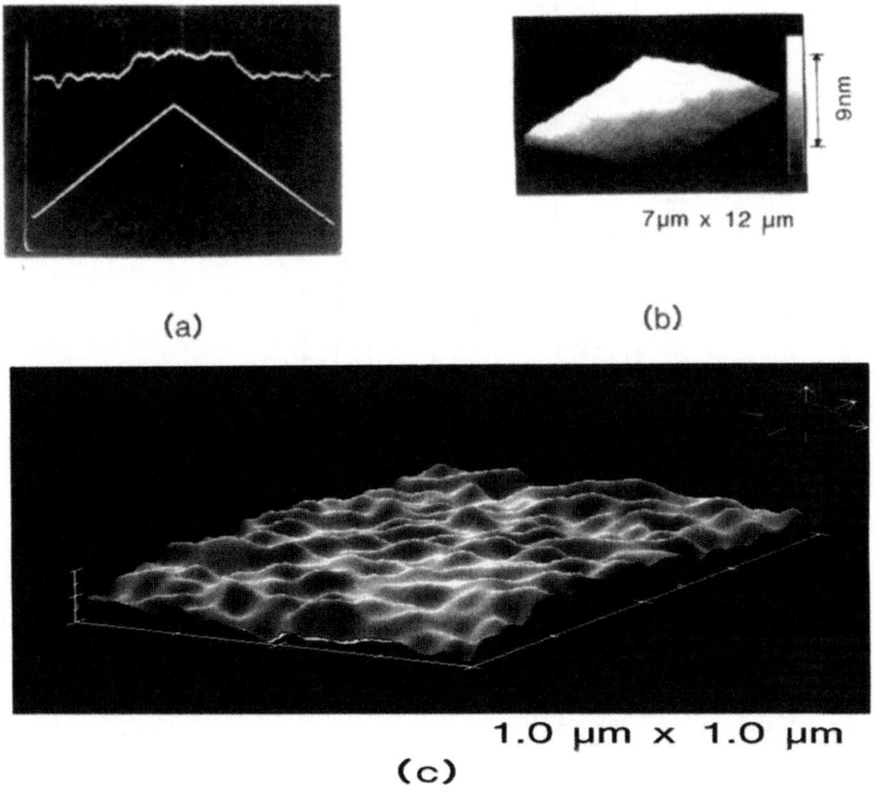

Fig. 7.7. Measured results by using a transmission PSTM (after Ohtsu, Nakagawa, Jiang, and Tomita [17]). (a) and (b) are for a 9 nm thick SiO_2 film which was coated on a half of a prism's surface. (a) Upper: Trace of an output signal from the system when the fiber was scanned along the x-axis on the sample. Lower: Saw-tooth waveform of the voltage applied to the PZT for scanning the fiber. (b) Three dimensional image of the sample obtained by scanning the fiber along the x- and y-axes. (c) Three dimensional image of the latex particles of 80 nm diameters.

By using a multi-core fiber [19], one may be able to measure the dynamic properties of the stimulus propagation on the biomembrane. In order to fully analyze the performances of the PSTM, the theory of near field wave optics should first be established. Furthermore, as a future ultimate technique, it may be expected that this sort of optical tunnel-tip will be used as a manipulator to control the motion of a single atom for crystal growth in a single-atom level [17]. It can be concluded that advanced PSTM will be achieved by fully by utilizing frequency controlled semiconductor lasers.

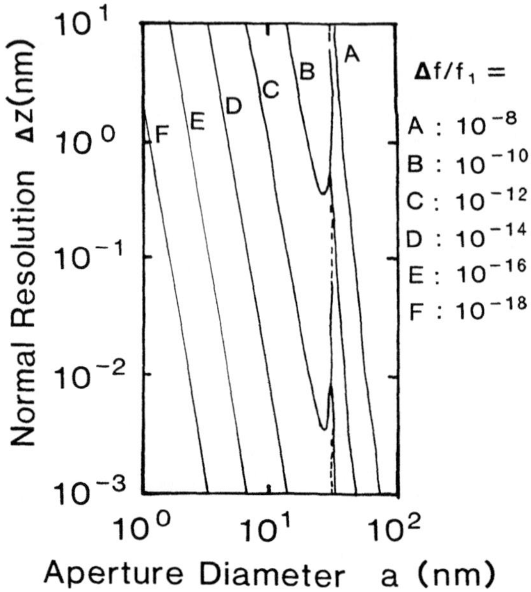

Fig. 7.8. Calculated relation between the aperture diameter and normal resolution of the reflection-resonance-type PSTM (after Ohtsu, Nakagawa, Jiang, and Tomita [17]). The parameter $\Delta f/f_1$ is the residual frequency fluctuations of the phase locked loop which is normalized to the optical frequency f_1.

7.3 Analytical Spectroscopy

(1) Laser radar (Lidar): Remote sensing of densities and locations of pollutant gases is possible by measuring the scattering or absorption of laser light. High power lasers, e.g., CO_2, YAG and Ar^+ lasers, have been used in conventional systems. Figure 7-9(a) shows an example of such a lidar system. A high frequency stability has been required for lasers to achieve a differential absorption lidar system by which the difference in absorption between two wavelengths is measured. To satisfy this requirement, the techniques described in chapter 6 have been utilized. In the case of a coherent lidar system, further FM noise reduction is required as was the case for coherent optical communication systems. Although the power in semiconductor lasers is lower than that in these lasers, their small size and low power consumption are advantageous features which can be used in practical air-borne or automobile-borne lidar systems. Furthermore, by utilizing the high efficiency of direct modulation of semiconductor lasers, a pseudo-random modulation technique has been employed to achieve high sensitivity and high spatial resolution in lidar systems. Field tests have confirmed that such pseudo-random modulation CW lidars showed a maximum measurable distance of 1 km for aerosols, and 3 ~ 5 km for clouds or dust with a spatial resolution of 9 m [20]. Figure

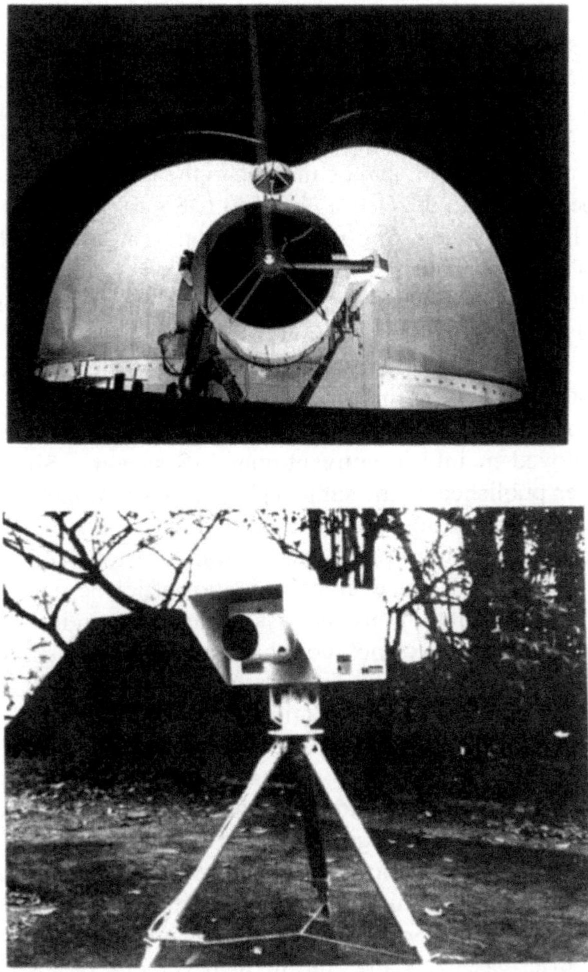

(a)

(b)

Fig.7.9. Laser radar station (a), and the transmitter head by using a semiconductor laser (b) (by the courtesy of N. Takeuchi, The National Institute for Environmental Studies, 1989).

7-9(b) shows a transmitter head by using a semiconductor laser. In the lidar system used to measure the ozone density in the stratosphere, a heterodyne technique between solar light and lead-salt compound semiconductor lasers (several μm ~ 10 μm wavelength) has been employed. As a result, the detection sensitivity as high as the shot noise level was achieved [21].

(2) Isotope separation: Lasers have been used for the isotope separation of Uranium (U) in laser fusion and of several atoms as tracers in medical diagnostics. Although the conventionally used gas-diffusion method falls into the category of

statistical separation process, the method using lasers is an individual separation process because the relevant isotopic atom or molecule is selectively excited by the lasers. This isotope can be completely separated from the others by using additional techniques, e.g., successive laser excitations; consequently, it can be claimed that the efficiency of laser isotope separation (LIS) is very high. Lasers should be highly stable in frequency and power, and be tunable to the resonance spectral line of the relevant isotope in order to develop a practical LIS system.

For the selective excitation of relevant isotopes, an isotope shift in resonant spectral lines of atoms or molecules is utilized. The principal causes of isotope shift in atoms are the mass, volume and nuclear-spin effects. The mass effect of molecules is larger than that of atoms and is the main cause in the isotope shift of molecules. To separate selectively excited isotopes, the techniques of photo-ionization, photo-dissociation, infrared multiple-photon dissociation and photo-chemical reaction, have been employed. Dye lasers or infrared molecular gas lasers have been employed mainly in conventional LIS systems. Since many review articles have been published on the subject [22], only a new method making use of semiconductor lasers is introduced in the following.

For the isotope separations of Rb and Cs atoms, the technique of laser induced drift has been proposed [23]. However, its separation efficiency is not yet sufficiently high owing to the de-excitation process associated with optical pumping. It was found that the separation efficiency could be increased by using semiconductor lasers. Since the direct FM response characteristics of semiconductor lasers exhibit a resonant peak at the relaxation oscillation frequency, a series of FM sidebands appear in the field spectrum of these lasers. By careful adjustment of the bias level, the frequency separation between the optical carrier and the FM sideband of the second-order can be made to coincide with the frequency separation between the two hyperfine energy levels of the ground states of these atoms. By simultaneously pumping the atoms with an optical carrier and with this FM sideband, a cyclic excitation was realized. Due to this cyclic excitation, for example, most of ^{87}Rb atoms can be maintained at excitation states so as to achieve a large cross section of the collision with buffer gases. As a result, the drift velocity of ^{87}Rb in buffer gases could become different from that of ^{85}Rb and consequently the separation efficiency has been improved [23]. For this method, the technique described in 6.3.1(a) has to be employed to lock the optical carrier frequency of the laser at the ^{87}Rb resonant frequency. As is shown in Fig. 7-10, a separation efficiency as high as that obtained by using conventionally used dye lasers (150 mW power) can be achieved by an AlGaAs laser with a power of only 3.3 mW [23]. In the isotope separation of Ba and Ga by using the Doppler-free two-photon transition process, one of the dye lasers has been replaced by a frequency controlled semiconductor laser [24]. For these purposes, and especially for the isotope separation of U, it is expected that the recently developed visible AlGaInP laser will be used as an efficient pumping source [25]. Its field spectral linewidth has been reduced to less than 50 kHz and its wavelength tuning range was expanded to 5 nm by using the optical feedback

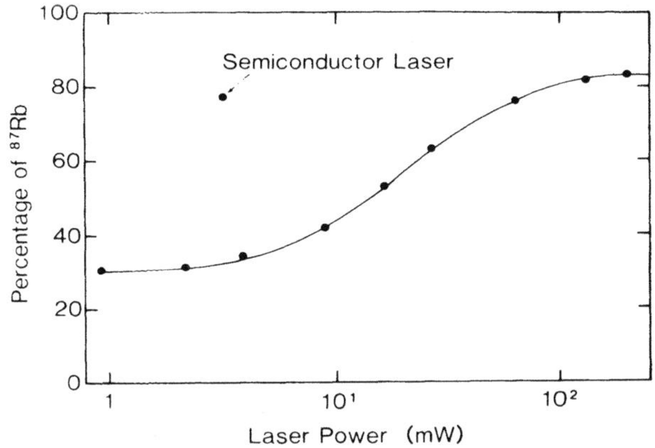

Fig. 7.10. Experimental result of the isotope separation of Rb using the method of laser induced drift (after Streatere, Mooibroek and Woerdman [23]). Closed and open circles represent the results by using semiconductor laser and dye laser, respectively.

technique shown in Fig. 6-11 [26]. Its power was amplified to 25 dB by using a dye solution and opto-galvanic spectral shapes of the D_1 and D_2 lines in Li have been observed with a very high signal-to-noise ratio [26].

(3) Analysis of SiH_x radicals: The fabrication of amorphous silicon film by chemical vapor deposition has been a key technology in the semiconductor device industry. For an accurate control of the thickness of a deposited film, quantitative analyses and the assignment of SiH_x species are required. For example, it has been confirmed by UV photolysis that an SiH_2 radical could emit fluorescence due to transitions between vibration energy levels. It can be expected that visible AlGaInP lasers will be used as light source for efficient and practical spectral analyses in UV photolysis systems [24]. This is because the emitted fluorescence spectra, due to the excitation by a pulsed ArF excimer laser, are distributed within a wavelength region of 0.6 μm, as shown in Fig. 7-11 [27].

7.4 Optical Pumping of Atomic Clocks

Stable frequency microwave oscillators have been used as key devices in microwave communication, broadcasting, navigation, earthquake prediction, astronomy, and so on. In these systems, atomic clocks have been used since they are the most reliable oscillators. Drastic improvements in the performance of these clocks have been attained by using semiconductor lasers as coherent optical pumping sources. Two examples of these improvements are described below.

(1) Cesium atomic clock at 9.2 GHz: A cesium (Cs) atomic clock is a microwave oscillator, for which the transition frequency (9.2 GHz) between two hyperfine levels ($F = 4$, $m_F = 0$ and $F = 3$, $m_F = 0$, see Fig. 7-12(a)) in the ground

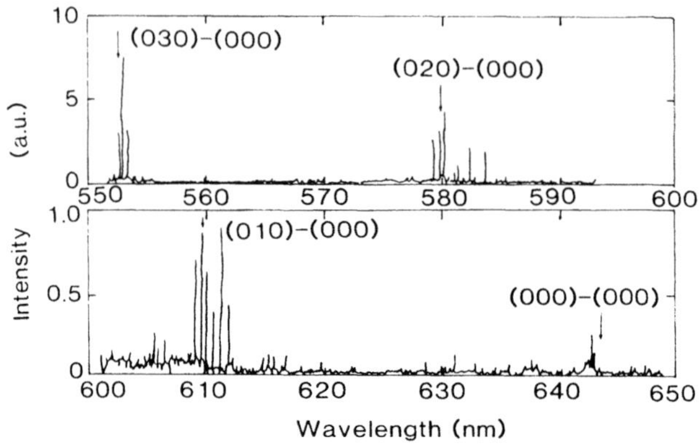

Fig. 7.11. Fluorescent spectral lines of SiH_2 radicals excited by an excimer laser (after Inoue and Suzuki [27]).

state of Cs has been used as the frequency reference to control the frequency of a voltage-controlled crystal oscillator (VCXO). Since the frequency accuracy of this clock is very high, it has been used as a primary standard of time. This clock has been studied and developed in a number of major research institutes in the world, and has been used as a time-keeping clock. A compact and portable Cs atomic clock has also been developed for satellite-borne systems.

In a conventional Cs atomic clock, the deflection of a Cs atomic beam by D.C. magnetic fields has been used to select atoms at the energy levels of $F = 4$, $m_F = 0$ and $F = 3$, $m_F = 0$. In such a state selection scheme, atoms at the $m_F \neq 0$ levels among the $2F + 1$ magnetic sublevels can not be used, which limits the efficiency of the state selection. Furthermore, this clock has revealed inherent sources of errors limiting its frequency accuracy, such as the frequency shift induced by magnetic fields which were used to deflect the atomic beam. To overcome these difficulties, an optical pumping scheme has been proposed [28]. Figure 7-12(b) shows a typical experimental setup in an optically pumped Cs atomic clock. Two 0.85 μm AlGaAs laser frequencies, ν_1 and ν_2, are tuned to transition frequencies from $F = 4$ to $F' = 4$ and from $F = 3$ to $F' = 4$, respectively, where F' represents the quantum number of the excited state. Following the cyclic transitions between ground and excited states, due to simultaneous optical pumping by the two lasers, all atoms are transferred to the $F = 4$, $m_F = 0$ level of the ground state. Therefore, the efficiency of the state selection is increased to eight times that of the conventional magnetic deflection scheme. Subsequent to this state selection, atoms pass through the two spatially isolated microwave cavities, from which a fringe-shaped microwave Ramsey spectral profile can be observed. Figure 7-13 shows a typical fringe-shaped Ramsey spectral profile [29]. (The occurrence of this fringe is due to the same principle the optical Ramsey fringe shown in 5.2.1(3).) The half linewidth of the center part of this fringe is very narrow and is determined by the inverse of the

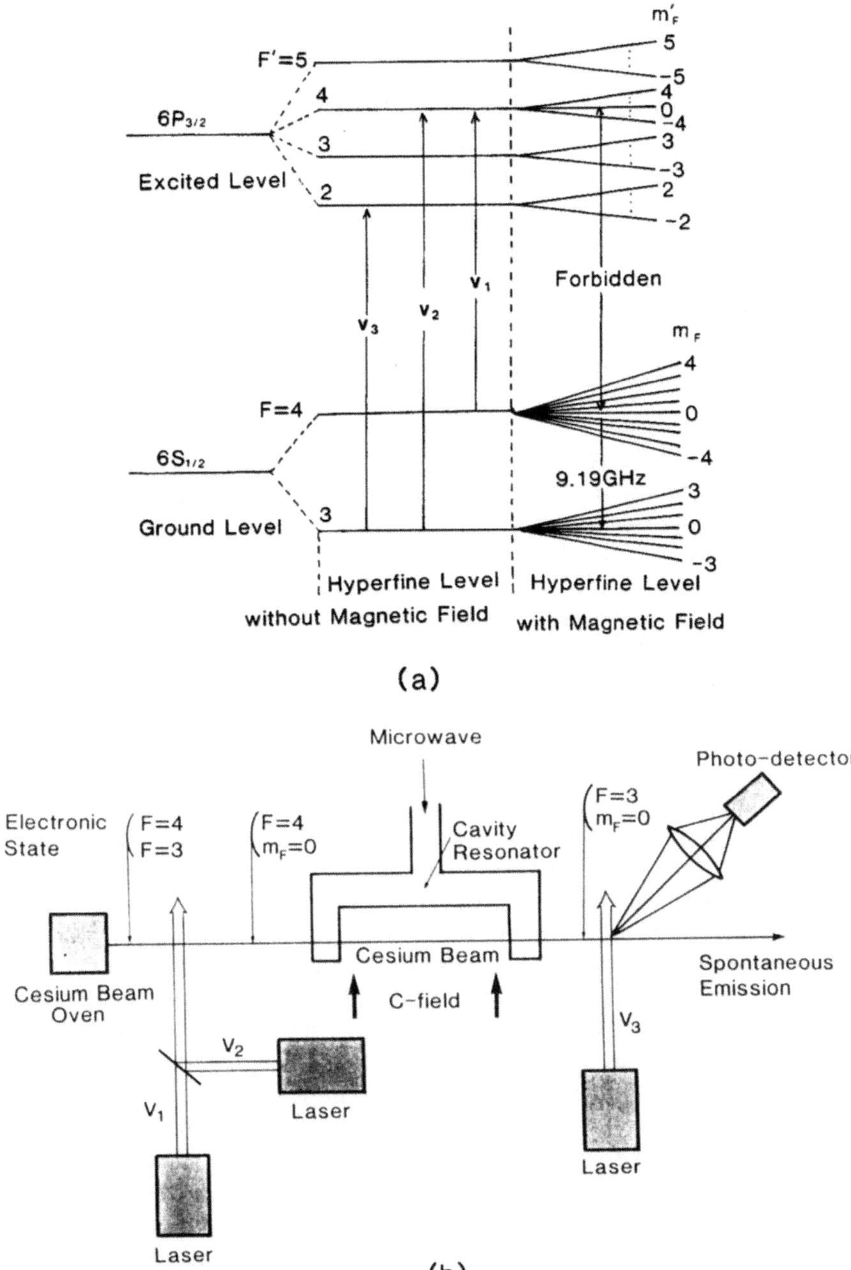

Fig. 7.12. A cesium atomic clock. (a) Cesium energy levels. ν_1 and ν_2 are the laser angular frequencies for pumping. ν_3 is for detecting the Ramsey fringe. (b) Block diagram of the experimental setup.

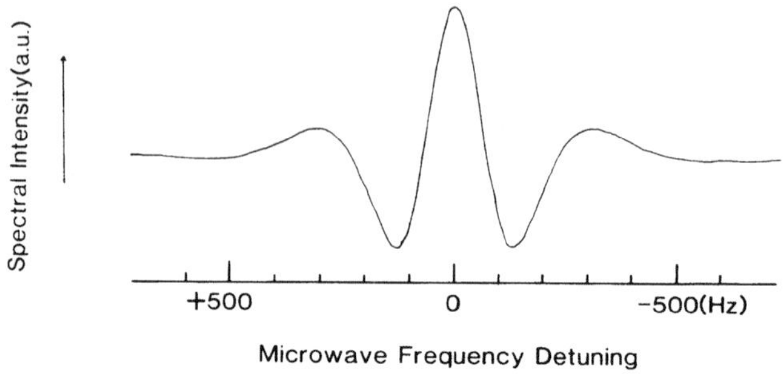

Fig. 7.13. Spectral lineshape of the microwave Ramsey fringe of a Cs atomic clock pumped by a semiconductor laser (after Ohshima, Nakadan and Koga [29], ©1988 IEEE).

separation between the two microwave cavities. This narrow spectral line is used as a sensitive and stable frequency demodulator to control the microwave frequency. Since Eq.(5-4) shows that a narrower linewidth is more advantageously used as a frequency demodulator, the Ramsey resonance method has been favorably employed [30].

The third AlGaAs laser is used to measure the Ramsey spectral shape. Its frequency ν_3 is tuned to the transition frequency between $F = 3$ and $F' = 2$ in order to excite the atoms at the $F = 3$ level which have been de-excited from the $F = 4$ level due to the microwave transition. The Ramsey spectral shape can be measured by detecting the fluorescence from the $F' = 2$ level which is emitted by the optical pumping from $F = 3$ to $F' = 2$ of this laser.

By using these techniques, i.e. state selection by two lasers and Ramsey spectral measurements by third lasers, most problems associated with the magnetic deflection scheme can be solved. Although an optical pumping scheme has been proposed for over ten years [28], experiments carried out with this scheme have progressed slowly owing to the lack of frequency controlled AlGaAs lasers with an 852 nm wavelength. However, rapid progress in developing stable AlGaAs lasers, as was described in 6.3.1(a) ~ (c), has recently led to the drastic improvement of this scheme [31]. An example of experimental setup is shown in Fig. 7-14 [32]. By using optical pumping, it can be expected that a frequency accuracy higher than 5×10^{-14} (which is the theoretical limit of conventional Cs atomic clocks) will be achieved in the near future. For the further improvement of this optically pumped Cs atomic clock, the usage of a laser cooling technique has been proposed [33].

(2) Rubidium atomic clock at 9.2 GHz: The rubidium (Rb) atomic clock has been popularly used as a compact and inexpensive microwave oscillator because its short-term frequency stability is higher than that of Cs atomic clocks, even though its frequency accuracy is lower. As shown in Fig. 7-15(a), the transition frequency (a Bohr frequency of 6.8 GHz) between the two hyperfine levels in the ground state ($F = 2$, $m_F = 0$ and $F = 1$, $m_F = 0$) is used as the reference frequency to control the VCXO.

Fig. 7.14. View of a Cs atomic clock pumped by a semiconductor laser (a picture from the National Research Laboratory of Metrology [32]).

In order to use this transition spectral line as a frequency demodulator, its spectral profile should be detected with a high sensitivity. For this purpose, atoms in the $F = 1$ level of the ground state are pumped by irradiating the light which is resonant to the transition from the ground to the excited state. The excited atoms immediately decay to the $F = 1$ or $F = 2$ level of the ground state. As a result, the population of the $F = 2$ level of the ground state increases while that of the $F = 1$ level decreases, and finally, in the steady state, the population difference between these two states approaches a constant value. However, by irradiating the microwave resonant to the Bohr frequency between these levels, this population difference is varied by the stimulated emission and absorption of microwaves. This variation also changes the absorption coefficient of the pumping light. Thus, the microwave transition spectral profile can be measured by measuring the light intensity transmitted through the Rb vapor cell as a function of the microwave frequency. This technique of resonating the light and microwave simultaneously to the atoms is called optical-microwave double resonance. By this technique, the measurement sensitivity of the microwave resonance spectral profile is increased. This is because the microwave transition is measured by monitoring the optical transition, which an energy amplification of 10^5 corresponding to the frequency ratio of the microwave and the light.

As was shown by Eq.(5-4), the double resonance spectral width should be narrow in order to detect accurately the FM noise of the microwave. Buffer gases, e.g., Ne, Ar and N_2, have been used to fill the Rb vapor cell to reduce the spectral width generated by the atomic collisions between Rb and these buffer gas atoms. As a result of this collisional narrowing, a spectral width as narrow as 100–500 Hz has been obtained.

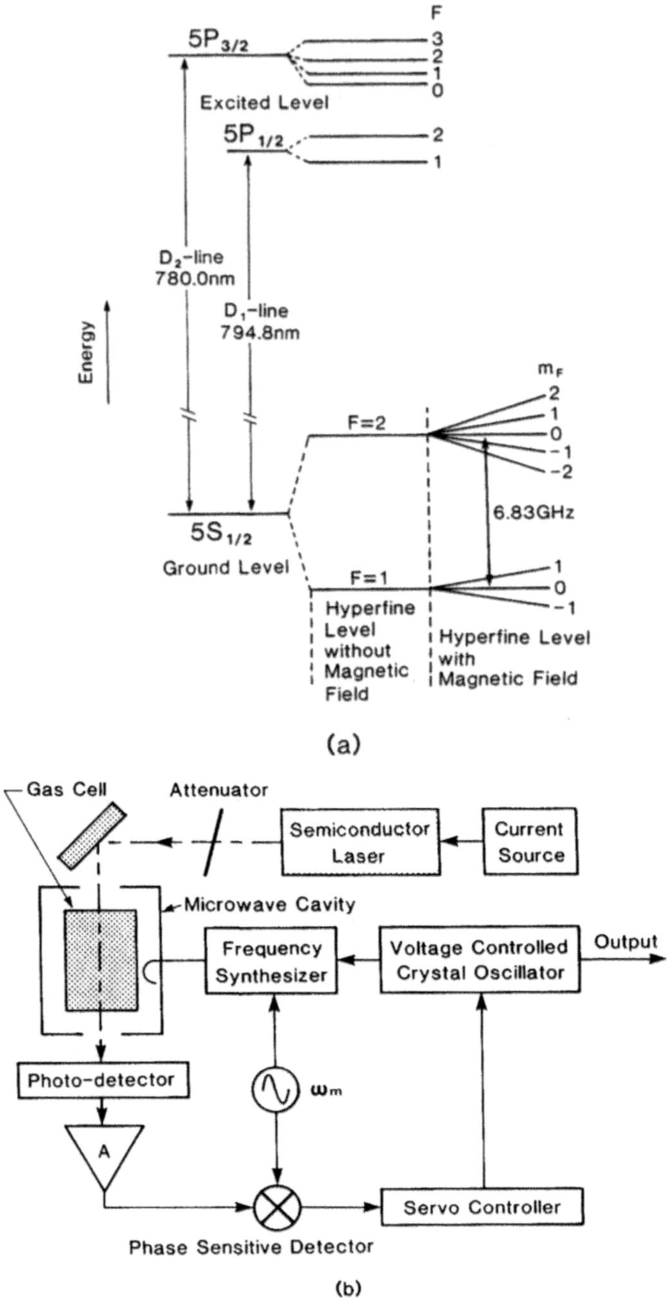

Fig. 7.15. A Rb atomic clock. (a) Energy levels. (b) Block diagram of the experimental setup.

Conventional Rb atomic clocks have used an incoherent discharge lamp for optical pumping. However, the center frequency of the double resonance spectral profile could be shifted due to variations in lamp power. This shift, called the "light shift", is attributed to the A.C. Stark effect produced by an optical field. This light shift has limited the long-term frequency stability of Rb atomic clocks. Furthermore, the frequency accuracy has also been limited by the frequency shift induced by collisions between Rb and buffer gas atoms. For these reasons, the Rb atomic clock has been considered as the secondary standard of time.

The Rb atomic clocks have been installed on board satellites in the global positioning system (GPS) to improve the system performances. For this purpose, the optical pumping scheme by using a 780 nm wavelength AlGaAs laser has been proposed. Its system diagram is shown in Fig. 7-15(b). In this scheme, using a low FM noise semiconductor laser, the effect of microwave frequency modulation could be transferred to the optical field of the laser in the Rb vapor cell. This modulation transfer occurs due to the nonlinear susceptibility of the Rb atoms which interact with the optical and microwave fields simultaneously. Since the double resonance spectral signal is detected by using this modulated laser light, the effect of this modulation transfer induces some sort of interference fringe on the double resonance spectral profile, as is shown in Fig. 7-16(a) [34]. The width of the center part of this fringe-shaped spectral profile was about 0.05 times that of a conventional lamp-pumped Rb atomic clock. The principle behind obtaining such a fringe-shaped spectral profile is equivalent to that in FM laser spectroscopy [35]. As is shown in Fig. 7-16(b), such a specific spectral profile could be accurately reproduced by using the theoretical model based on the density matrix formalism of three-level atoms. It has also been confirmed that the fringe-shaped spectral profile of Fig. 7-16 is equivalent to the fringe-shaped Ramsey spectral profile in a Cs atomic clock (see Fig. 7-13). In the case of Cs atomic clocks, perturbations from microwave fields were applied to Cs atoms at two separate positions along the spatial axis of the Cs atomic beam trajectory. On the other hand, in Rb atomic clocks, perturbations from optical fields were applied to Rb atoms at separate positions along the optical frequency axis as a result of modulation transfer. The effect of these separated perturbations induced a fringe on the double resonance spectral shape. By using this narrow linewidth spectral shape as a frequency demodulator, it was confirmed that the short-term frequency stability was improved 40 times over that of conventional Rb atomic clocks [36, 37].

Further improvements of short-term frequency stability were confirmed by reducing the FM noise of the laser at the Fourier frequency of the microwave modulation frequency since the noise contained in the double resonance was originated mainly from the FM noise of the laser [36]. By reducing this FM noise down to the shot noise limit of the laser frequency control loop, it was estimated that the short-term stability of the Rb atomic clock could become 1,000 times higher than that of the presently developed laser-pumped Rb atomic clock [36, 38].

The magnitude of the induced light shift can be quantitatively evaluated by

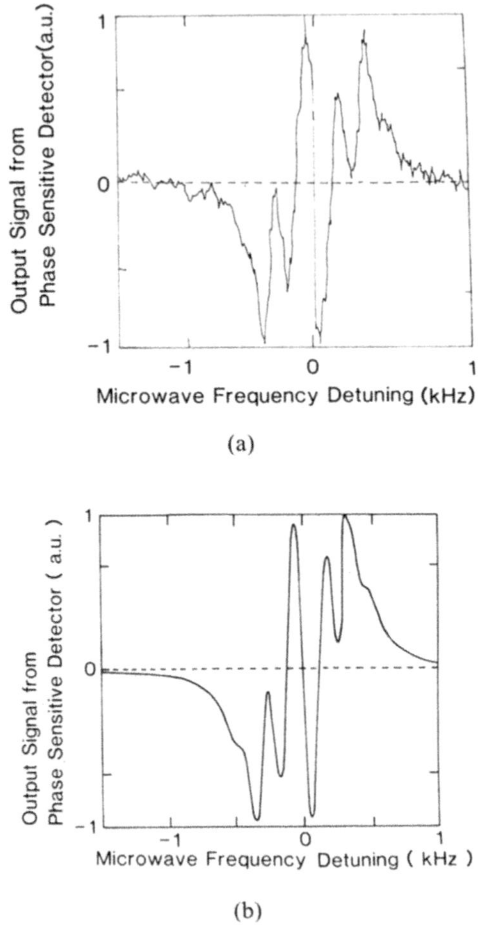

Fig. 7.16. Derivative of an optical-microwave double resonance spectral shape of a Rb atomic clock pumped by a semiconductor laser (after Hashimoto and Ohtsu [34], ©1987 IEEE). (a) Measurement result. The width at the center of the curve (corresponding to the interval between points A and B is about 1/20 times that of pumped by a lamp. (b) Calculated result based on the density matrix.

varying the laser power and frequency; this is shown in Fig. 7-17(a) [34]. A negative electrical feedback system was developed to control the injection current of the laser in order to suppress the light shift of the microwave frequency. As a result of this self-tuned system, the drift in microwave frequency has been reduced; this result is shown in Fig. 7-17(b) [37], from which it can be confirmed that the long-term frequency stability was improved by 40 times over that of conventional ones. This stability could be even higher than that in commercially available portable Cs

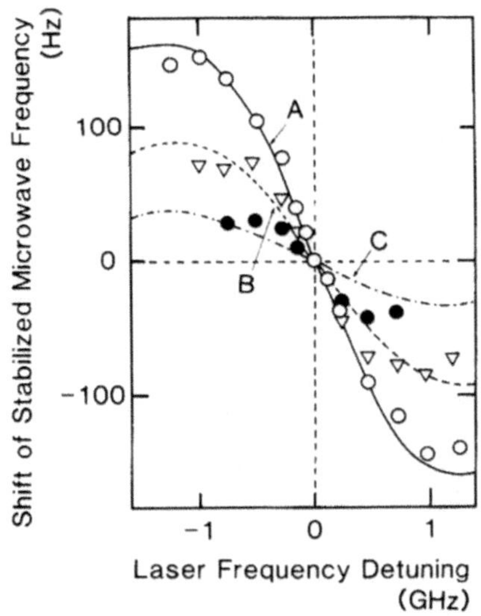

(a)

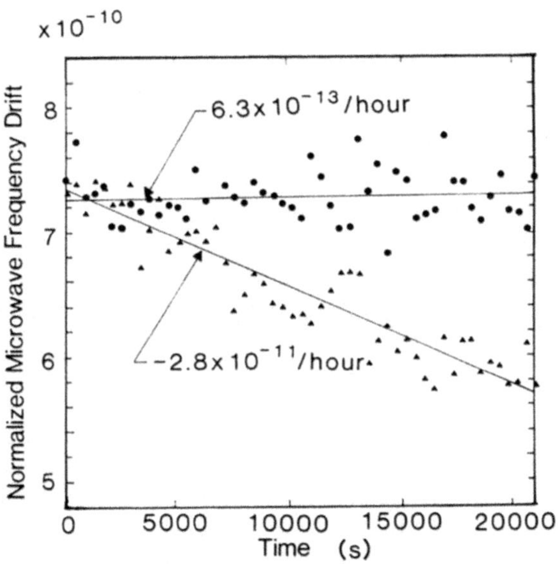

(b)

Fig. 7.17. Light shift and its suppression of the microwave frequency in a semiconductor laser pumped Rb atomic clock. (a) Magnitude of the light shift. Curves A, B and C show the results when laser power densities are 1.00 μw/cm^2, 0.36 μw/cm^2 and 0.14 μw/cm^2 respectively (after Hashimoto and Ohtsu [34], ©1987 IEEE). (b) Result of the drift suppression of the microwave frequency in the self-tuned system (after Hashimoto and Ohtsu [37], ©1990 IEEE). Closed and open circles show the results in the cases of with and without the self-tuned system, respectively.

atomic clocks [36, 37, 38]. Replacement of an Rb vapor cell by an Rb atomic beam has been proposed to improve the frequency accuracy and to eliminate the atomic collisions with buffer atoms [39]. This is expected to establish a new primary standard of time due to the improvement of accuracy.

As a result of the intensive studies described above, it is expected that a novel, high-performance, compact and low price Rb atomic clock will be developed in the near future, and will be, thus, incorporated in key application systems such as GPS, ISDN (Integrated Services Digital Network) and in a new primary standard of time.

7.5 Quantum Optics and Basic Physics

(1) High resolution spectroscopy of atoms and molecules: High resolution laser spectroscopy could provide accurate information on the structures and dynamic properties of the excited states, and so on, of atoms and molecules. The frequency demodulators presented in 5.2.1(1) ~ (6) have made use of the results of high resolution spectroscopy obtained by using several low FM noise lasers. That is, low FM noise lasers are required to observe the narrow linewidth spectral profiles in atoms and molecules; furthermore, FM noise reduction is possible by using these spectral profiles as a high quality frequency demodulator. Thus, laser spectroscopy and frequency control are mutually related with each other.

Wavelength tunable dye lasers have been used for conventional laser spectroscopy in the visible region. However, high resolution spectroscopy in the near infrared region (wavelength of 0.7 ~ 1.6 μm) has been actively carried out recently by using greatly improved semiconductor lasers. The advantage of the latter type of spectroscopy is the possibility of obtaining a wideband frequency sweep by varying the laser temperature and/or injection current. In addition, improvement in the sensitivity of these spectrometers can be expected by employing the technique of FM spectroscopy [40], which is possible by making use of the high FM efficiency of semiconductor lasers. A great number of Doppler-broadened spectral lines due to the higher harmonics and combination tones of vibration-rotation transitions in organic molecular vapors can be observed within a wavelength range of 0.7 ~ 1.6 μm. Although their absorption is weak, spectral lines caused by $2\nu_1$, $2\nu_3$ or $\nu_1 + \nu_3$ vibration transitions in NH_3 and $2\nu_2 + \nu_3$ vibration transitions in H_2O have been observed at a wavelength of 1.5 μm with a detection sensitivity of 2.3×10^{-3} Torr per unit length [41]. Furthermore, those caused by the $2\nu_1 + \nu_2 + \nu_3$ vibration transition in H_2O have been measured at a wavelength of 0.8 μm [42]. In that case, fairly strong absorptions were observed because the $2\nu_1 + \nu_2 + \nu_3$ band is coupled with the $\nu_2 + \nu_3$ band by Darling-Dennison resonance [43]. The wavelengths of these absorption lines have been calibrated within an error of $1 \times 10^{-7} \sim 1 \times 10^{-8}$. Following these pioneering works, spectroscopic data accumulated for several organic molecular vapors; for example, spectral lines in C_2H_2 [44] were obtained by sweeping a DFB laser frequency over 1.5 THz. Furthermore, the high resolution spectroscopy of second harmonic vibration transitions in HCN [45] and NH_3 [46]

was also obtained by DFB lasers.

A highly sensitive spectrometer for the wavelength regions of 0.75 ~ 0.88 μm, 1.3 μm and 1.5 μm has recently been constructed [46], and the absorption measurement sensitivity of 2×10^{-6} has been achieved. By using this spectrometer, a systematic study of the characteristics of spectral measurements of the vibration transitions in NH_3 has been carried out. Results show that the transfer from normal mode vibration to local mode vibration is found with an increase in the vibration quantum number. These experimental data were compared with theoretical analyses and were in good agreement. Further improvements in the sensitivity of these spectrometers can be expected due to the technique of frequency modulation (FM) spectroscopy [47]. Several experimental results in FM spectroscopy have already been demonstrated [40]. The problem remaining to be solved in FM spectroscopy by semiconductor lasers is the fact that the intensity modulation (IM) is simultaneously induced, which limits the measurement sensitivity. However, it has been proposed that the contribution from the IM could be reduced by making use of the two-tone FM technique. As a result, its sensitivity has been improved to ten times that of the conventional one [48].

Spectral lines due to electronic transitions in atomic vapors have been measured in the wavelength region of 0.8 μm. Systematic studies on Rb [49] and Cs [50] have been reported. For inert gases, the technique of opto-galvanic spectroscopy has been employed because a great numbers of opto-galvanic spectral lines are distributed between the near-infrared and the visible region [51]. Although sufficiently reproducible frequency measurements could be rather difficult to obtain owing to the frequency shift caused by plasma instability and the Stark effect, a fairly high sensitivity in measurements has been obtained [52].

Since most spectral lines caused by electronic transitions in atomic vapors are distributed within the visible region, short wavelength lasers could be required for these measurements. For this purpose, a second harmonic wave generated from the active layer of an AlGaAs laser has been used to measure the strong spectral lines in K (transitions $5p^2P_{1/2}$–$4s^2S_{1/2}$ and $5p^2P_{3/2}$–$4s^2S_{1/2}$) and Al (transition $3p^2P_{3/2}$–$4s^2S_{1/2}$) vapors at a wavelength range of 0.4 μm [53]. As shown in Fig. 7-18(a), similar measurements of Rb lines, by using the second harmonic wave generated internally from a 1.56 μm InGaAsP laser, have also been obtained [54].

The recently developed 0.67 μm wavelength visible AlGaInP laser can be also used for laser spectroscopy of Li atoms. Figure 7-19 shows the opto-galvanic spectral profiles of the D_1 and D_2 lines in Li. By increasing the discharge current of the hollow cathode lamp, a spectral splitting can be clearly seen (especially, in the D_2 line of curve C), which is due to the D.C. Stark effect of Li induced by a strong electric field in the hollow cathode [26, 55].

Doppler-free spectroscopy of atomic vapors has been carried out by using the technique of atomic beam or nonlinear spectroscopy. Atomic beam spectroscopy has been carried out, e.g., for Cs, Rb, etc. with a high sensitivity. An example of sensitive experimental result for the Rb atomic beam is shown in Fig. 7-18(b) [39].

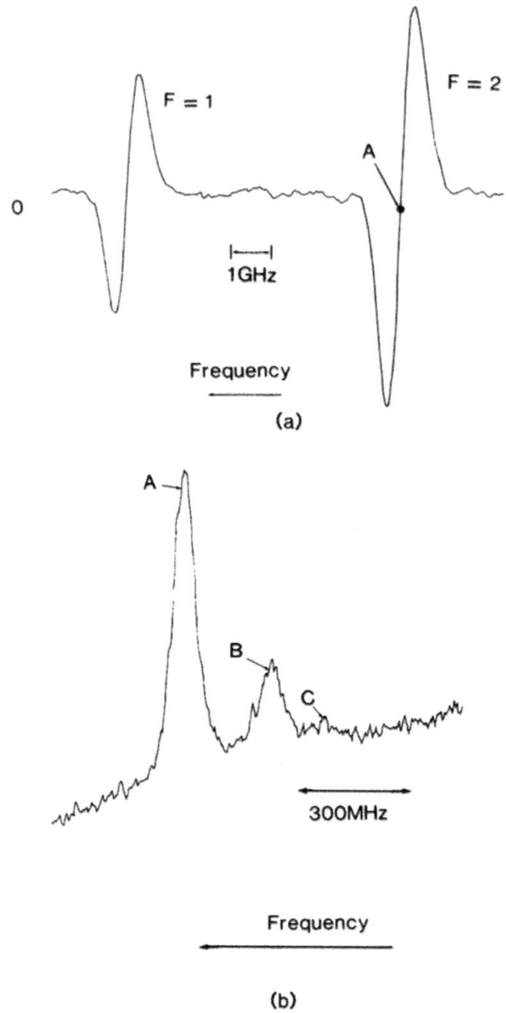

Fig. 7.18. Rb atomic spectral shape measured by a semiconductor laser (a) Derivative of Rb atomic absorption spectral shape measured by the second harmonic radiation in a 1.56 μm InGaAs laser (after Ohtsu and Ikegami [54]). A and B correspond to the transitions from $F = 1$ and $F = 2$ in the ground level to the excited level, respectively. (b) Fluorescent spectral shape of a Rb atomic beam measured by a 0.78 μm AlGaAs laser (after Furuta, Suzuki, and Ohtsu [39]). A, B and C correspond to the transitions from $F' = 3, 2, 1$ in the excited level to $F = 2$ in the ground level, respectively.

It was obtained by using a 0.78 μm AlGaAs laser and the half linewidth of the fluorescence spectral profiles caused by transitions from $F' = 1, 2,$ and 3 levels of the excited state to the $F = 2$ level of the ground state was as narrow as 70 MHz. This

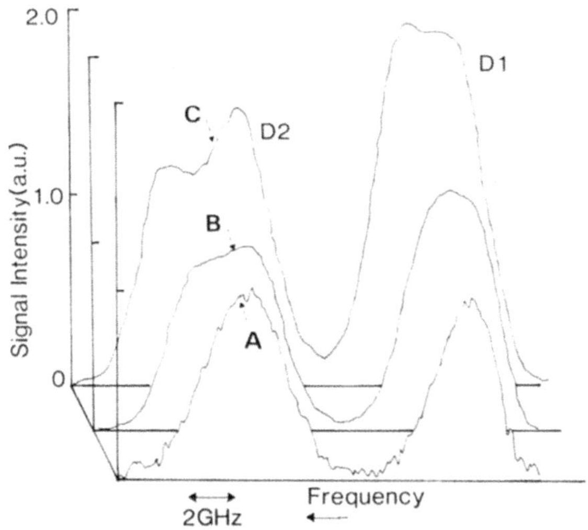

Fig. 7.19. Opto-galvanic spectral shapes of D_1 and D_2 lines in Li vapor (after Ohtsu, Suzuki, Nemoto, and Teramachi [26]). The discharge current for the hollow cathode lamp was 4(A), 7(B), and 10(C) mA.

linewidth was determined by the residual Doppler broadening due to a 5 degrees divergence of the atomic beam [39]. However, it is narrow enough so that each spectral component is clearly resolved.

As a first example of nonlinear spectroscopy, the saturated absorption technique has been employed, e.g., for Cs [50], Rb [49, 56] and so on. Figure 7-20(a) shows derivatives of the saturated absorption lineshapes in ^{87}Rb-D_2 obtained by using a 0.78 μm AlGaAs laser, in which eleven narrow spectral lines including cross-resonance lines, are resolved [56]. They have linewidths as narrow as 40 MHz, which is limited by the lifetime of relevant energy levels. Their spectral strengths and positions on the abscissa agree well with those estimated by theoretical analyses (see Fig. 7-20(b)).

As a second example, optical-optical double resonance has been carried out for the D_1 [57] and D_2 [58] lines in Rb by using 0.79 μm and 0.78 μm AlGaAs lasers, respectively. Figure 7-21 shows the double resonance spectral shapes for the D_2 component of an ^{87}Rb atomic beam [39], which exhibit clear Doppler-free spectral shapes. Because the pump laser frequency was fixed to the transition frequency between the $F = 2$ and $F' = 3$ levels, the strength of the spectral component $F' = 3$ is weaker than the other two components as a result of saturation. Figure 7-22(a) shows the new experimental setup for nonlinear pump-probe spectroscopy [58]. Doppler-free spectral profiles caused by the pumping saturation from a 0.78 μm AlGaAs laser were probed by using the second harmonic wave generated from a 1.56 μm InGaAsP laser active waveguide. Figure 7-22(b) shows the experimental result of the first derivative of the Doppler-free spectral shape. This spectral shape

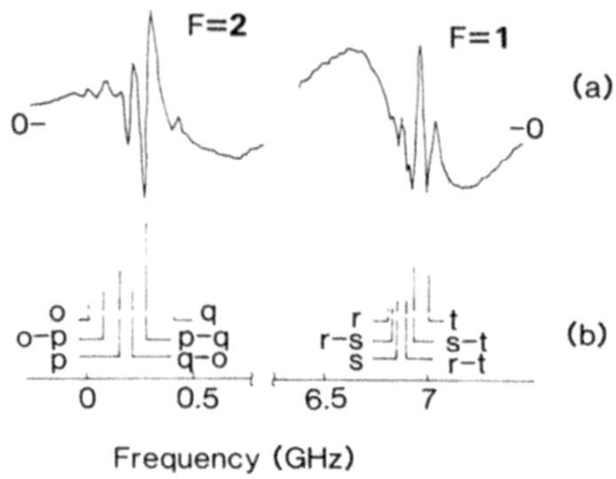

Fig. 7.20. Derivative of saturated absorption spectral shapes in ^{87}Rb (after Ohtsu, Hashimoto, and Ozawa [56], ©1985 IEEE). (a) Experimental results. (b) Assigned spectral lines. For the transition from the $F = 1$ level of the ground state, two saturated absorption lines (s and t) and three cross-resonance lines (r-s, r-t, and s-t) were observed. For the transition from $F = 2$ level, three saturated absorption lines (o, p, and q) and three cross-resonance lines (o-p, q-o, p-q) were robserved.

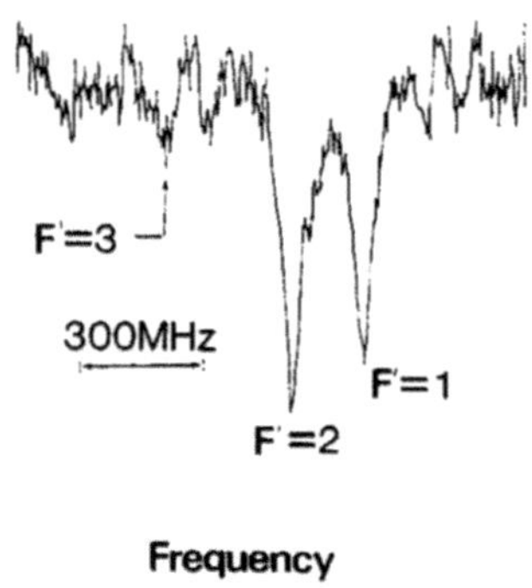

Fig. 7.21. Optical-optical double-resonance spectral shapes of ^{87}Rb atomic beam (after Furuta, Suzuki, and Ohtsu [39]). The pump laser frequency was fixed to the transition frequency from $F = 2$ to $F' = 3$.

can be used as a frequency reference to stabilize the frequency of pump- and probe-lasers simultaneously; thus achieving a stabilized frequency link between the 0.78 μm AlGaAs and 1.56 μm InGaAsP lasers. Similar results can be expected for Li atoms by using 0.67 μm AlGaInP and 1.34 μm InGaAsP lasers.

The third example is the Doppler-free two-photon spectroscopy for Li atomic vapor, which has been carried out by using a dye laser and an AlGaAs laser [59]. By using a thermoionic heat pipe diode as the sensitive detector for spectral measurements, clear spectral profiles have been obtained and the hyperfine components in ^{6}Li and ^{7}Li have been clearly resolved.

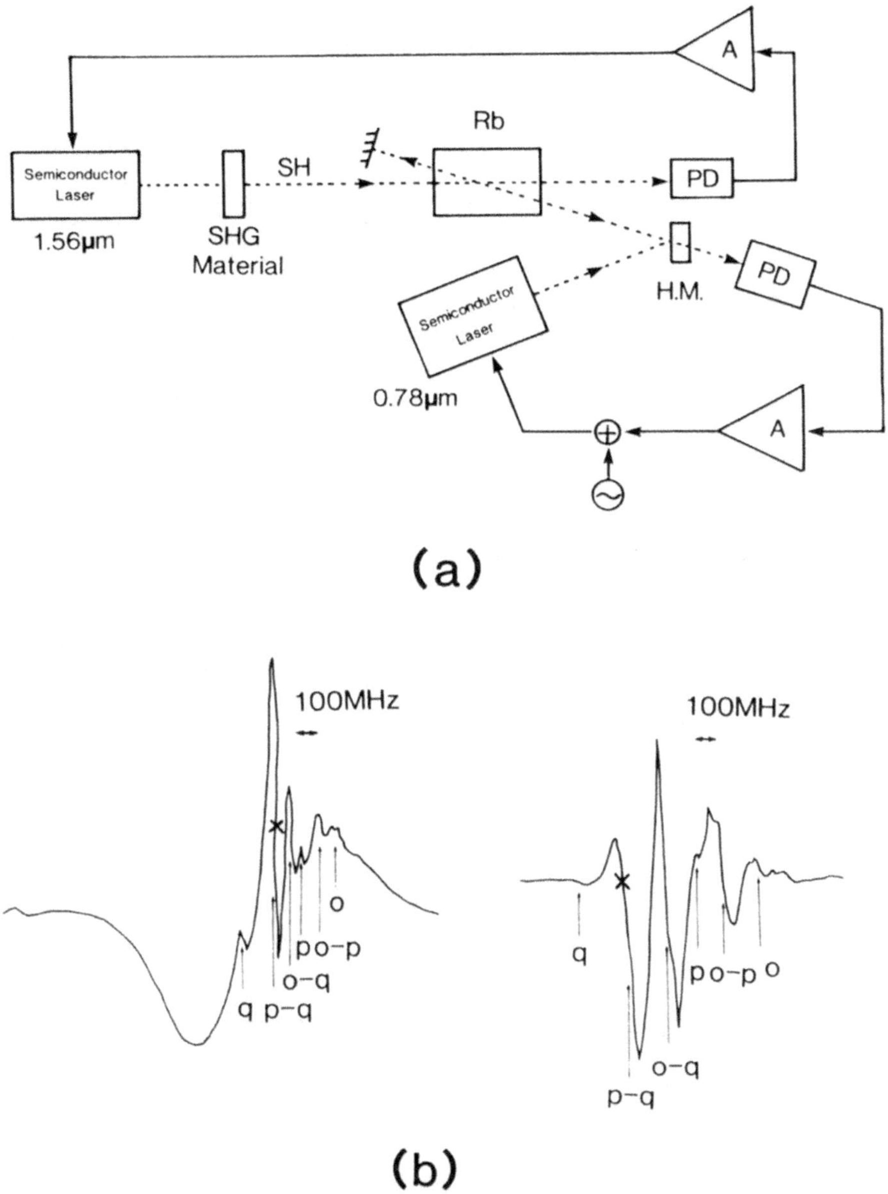

Fig. 7.22. Novel Doppler-free pump-probe spectroscopy of ^{87}Rb by using an AlGaAs laser and the second harmonics of the InGaAsP laser as the pumping and probe light sources, respectively (after Ikegami, Kusuzawa, Furuta, and Ohtsu [58]). (a) Experimental setup. (b) Measured profiles of the derivative of the saturated absorptions and cross-resonance lines. Transition was from $F = 2$ of the ground state. Left: Conventional spectral profile measured by a probe beam from the AlGaAs laser. Right: Optical-optical double resonance signal measured by the second harmonics of the InGaAsP laser.

As one of the other applications of semiconductor lasers to alkali vapor spectroscopy, Rydberg states of alkali atoms can be produced by the stepwise excitation of Cs and Rb [60]. By using this method, information on highly excited states can be obtained [61]. The Rydberg atoms prepared by this method have also been used in one atom maser in the microwave region, i.e. in one branch of the cavity QED.

Furthermore, as was described in 3.3, semiconductor laser-pumped tunable solid state lasers have been developed. Further improvement of the spectral measurement system can be expected if these low noise solid state lasers are used.

(2) Test of parity non-conservation and related studies: The existence of a weak interaction between elementary particles in atoms has been predicted by the standard electroweak theory developed by Glashow, Weinberg and Salam [62]. Experimental tests for this theory have been carried out in the field of high-energy physics by using a huge accelerator. However, it has been found that the method of precision laser spectroscopy could be used as a more convenient and less expensive tool for these tests [62, 63]. Since the parity conservation of the wave function of the atom could be violated in a weak interaction, the weak optical transition due to this violation can be detected by a carefully designed sensitive laser spectrometer. Although dye lasers have been conventionally used in these experiments, frequency controlled semiconductor lasers have also been used recently. Figure 7-23 shows an experimental setup to detect the change in polarization of the transmitted light through the Pb atomic vapor, in which a 1.3 μm InGaAsP laser is used as the coherent light source [64]. The optical transition caused by magnetic dipole interaction in Pb atoms was monitored to detect the change in polarization induced by

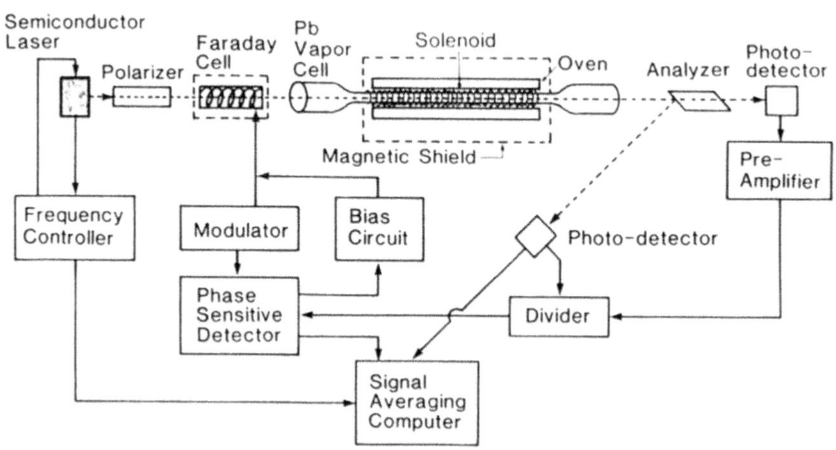

Fig. 7.23. Experimental setup to test parity non-conservation of Pb atoms (after Emmons, Reeves and Fortson [64]). Pb atomic vapor in an oven was irradiated by a 1.3 μm semiconductor laser whose wavelength is in resonance with the magnetic dipole of Pb atom, and the polarization of the transmitting laser beam was detected.

parity non-conservation. This sensitive spectrometer, employing the techniques of polarization modulation and phase sensitive detection, was sensitive to a polarization rotation of 0.1 micro radian. Stark coefficients in Cs atoms have also been accurately measured by using a carefully frequency-controlled AlGaAs laser, by which several structural constants of Cs have been estimated to test the parity non-conservation [65].

Semiconductor lasers can also be used to accurately measure the Lamb shift of muonic atoms [66]. A frequency controlled 0.73 μm AlGaAs laser is used as the master laser for the injection-locking of a high power solid state Alexandrite laser to reduce its FM noise. The third harmonic radiation from the locked Alexandrite laser is generated by using nonlinear optical media. This ultraviolet light can be used as a pumping source for muonic atoms.

(3) Manipulations of atoms and ions: The content of this sub-section has been partly described in 5.2.1(4) ~ (6). That is, experimental studies on the deceleration of atomic motion in vacuum, i.e. laser cooling [67], and confinement of atoms in a volume limited by laser beams have been recently developed owing to improvements of the laser frequency control technique. By irradiating the laser beam on the atoms from direction opposite to the trajectory of the atomic beam, photons with an energy lower than the energy difference between the two atomic energy levels are absorbed as a result of the first-order Doppler effect. The photon energy emitted spontaneously after this absorption is equal to the energy difference, and is larger than the absorbed photon energy. Therefore, by repeating these absorption and spontaneous emission processes, the atomic motion is decelerated by loosing the thermal energy of atoms. As a result of laser cooling technique, the relativistic phenomenon of the second-order Doppler effect, and thus, the shift in center frequency of the atomic spectral profile can be reduced, which can be advantageously used as a high quality frequency demodulator as described in 5.2.1(4). The lowest equivalent temperature of the cooled atom is lower than the value determined by the Doppler effect (<40 μK) [68, 69]. Furthermore, in a recent experiment, a temperature lower than 2 μK has been obtained which corresponds to a thermal velocity of 3 cm/s [70]. This is even lower than the value determined by recoiling between an atom and photon. To break through these limits, the technique using atomic badminton has been proposed [71]. Atomic badminton with cooled and slow atoms has been recently carried out by using a gradient-light force mirror, i.e. the gradient force of an evanescent wave from a Dove prism on which the cooled atom dropped [72].

It has become possible to confine a certain number of atoms (e.g., 1×10^7 atoms) with an atomic density as high as 1×10^{11} /cm^3 [73] to a limited volume in vacuum for more than several minutes. Such a high density confined atomic mass exhibits specific characteristics which are similar to those of viscous fluids; as result, this confined atomic mass has been called "optical molasses". Figure 7-24 shows the spatial profile of the optical molasses of Na atoms. For these experiments, dye lasers have been conventionally used. However, it has been recently recognized

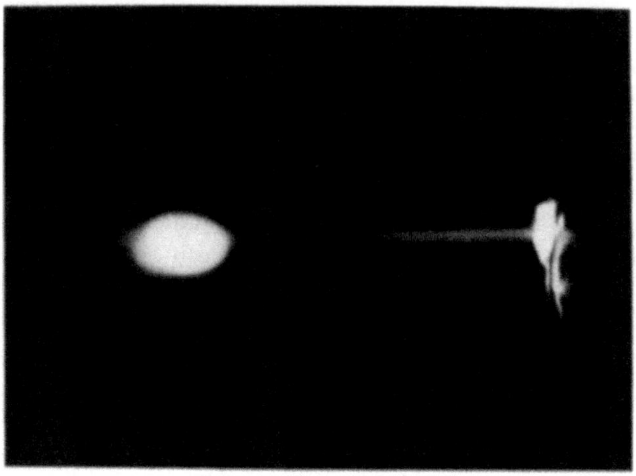

Fig.7.24. Optical molasses of Na atoms. The elliptic image is the optical molasses, and the bright part on the right is the nozzle to form the Na atomic beam (by the courtesy of S. Chu, AT&T Bell Labs., 1986).

that it is advantageous to use the high efficiency of frequency modulation and controllability of semiconductor lasers. Figure 7-25 shows an experimental setup using 0.85 μm AlGaAs lasers for optical pumping, laser cooling, and producing optical molasses of Cs atoms [74]. Laser coolings of Rb atoms [75] and rare gases [76] have also been carried out by using AlGaAs lasers. These experiments have

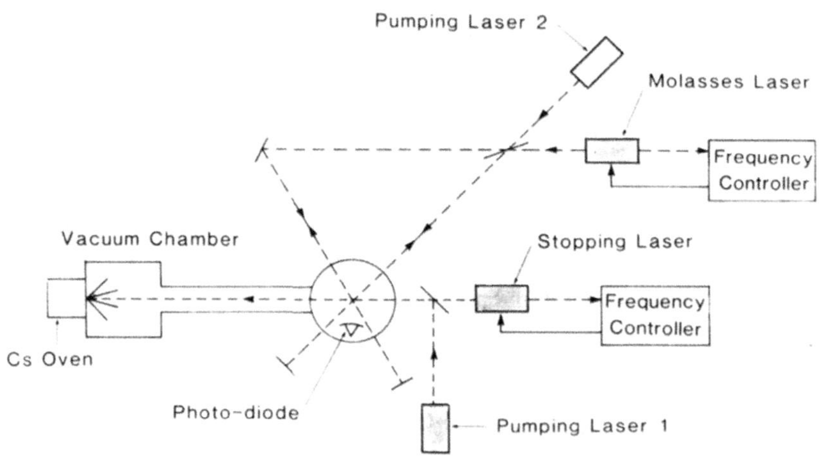

Fig. 7.25. Experimental setup of laser cooling for Cs atoms (after Sesko, Fan and Wieman [74]). By using several semiconductor lasers, deceleration, cooling, and furthermore, production of optical molasses were performed for Cs atomic beam.

been carried out by using expensive and complicated atomic beam apparatus. However, a simple experimental configuration using an inexpensive Cs vapor cell has recently been used to produce optical molasses [77].

As a method combining optical Ramsey and laser cooling, the atomic fountain method has been proposed [78], in which, as is shown in Fig. 7-26, an atomic beam is spouted upward. Only a single laser beam (#1) is required to generate optical Ramsey fringes. The atomic beam is irradiated by this laser beam from a transverse direction. If the atomic beam is simultaneously irradiated by the cooling laser (#2) from a longitudinal direction, the atomic beam is first decelerated and then falls down. Thus, the atoms cross the laser beam #1 twice, from which Ramsey fringes are observed. In this case, a very narrow linewidth Ramsey spectral shape can be obtained. A Ramsey fringe in the microwave frequency region was recently observed by using an Na atomic beam which was decelerated by laser cooling [79]. As shown in Fig. 7-27, its spectral width was as narrow as 2 Hz.

If these fringes were observed in the optical frequency region, this narrow spectral shape could be used as a high quality frequency discriminator as was described in 5.2.1. For example, since the transition spectral linewidth in Na is as narrow as 40 kHz when it is cooled down to several μK, the field spectral linewidth of the laser used for these cooling experiments should be narrower than 40 kHz. Furthermore, the laser frequency should be swept accurately over 1 GHz, for which the technique in 6.3.1(c) will be required.

The experimental study on ion trap, i.e. confining a single ion to the limited volume of an electro-magnetic potential, advanced remarkably by simultaneously making use of the laser cooling technique. Even in a cooled atomic beam, a shift in the center angular frequency ν_r of their resonant spectral shape could not be

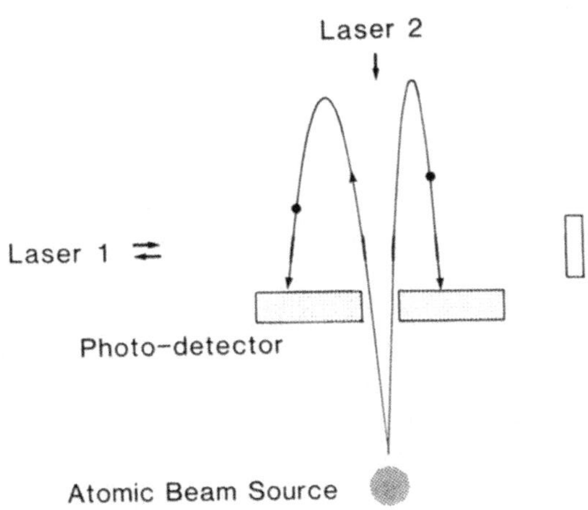

Fig. 7.26. Schematic explanation of the atomic fountain (after Beausoleil and Hänsch [78]). Laser #1 is for producing the Ramsey fringe, and laser #2 is for laser cooling.

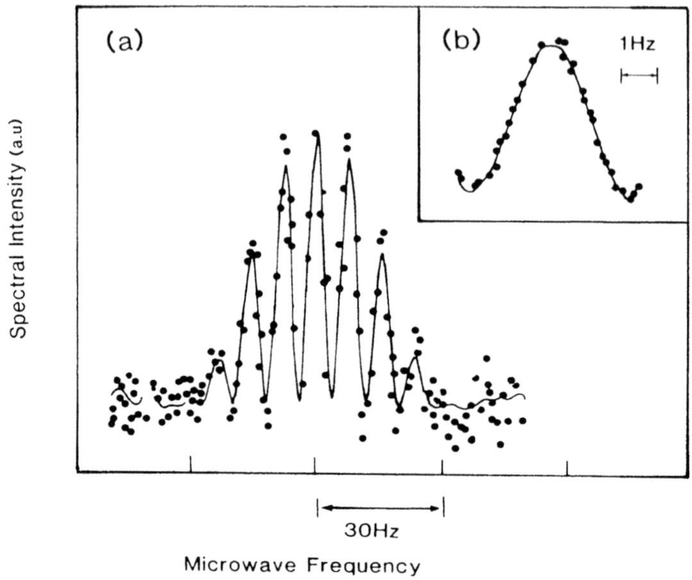

Fig. 7.27. Microwave Ramsey fringe obtained with an atomic fountain (after Kasevich, Riis, Chu, and De Voe [79]). (a) The whole profile. (b) Magnification of the central part of the profile.

neglected due to the collisions between cooled and confined atoms. To eliminate this shift, a single atom should be prepared in vacuum. If this atom is stopped by laser cooling, the second-order Doppler effect is also eliminated. For this purpose, it is more advantageous to use ions than neutral atoms because a deep electromagnetic trap is used for ions. A Penning trap composed of D.C. electric and magnetic fields, or a Paul trap composed of a D.C. electric and RF magnetic fields has been employed. Figure 7-28 shows the experimental setup in which an ion is injected into the electro-magnetic trap by the electron gun and trapped at the minimum of the electro-magnetic potential as a result of laser cooling [80]. Consequently, an ionic energy state, almost completely isolated from its surrounding circumstances, is achieved. This ionic state can be used as an ideal frequency discriminator in 5.2.1.

Several specific characteristics, e.g., quantum jump (a phenomenon in which an atom jumps from one energy level to another at the moment of absorbing or emitting a photon [81]), has been observed for a cooled and confined single ion [82, 83, 84]. Although this phenomenon was first discussed by N. Bohr in 1913, its experimental confirmation was not finalized until just recently. As is shown by Fig. 7-29, the switching signal caused by the quantum jump can be clearly observed by counting the number of photons emitted from the ion [83].

As a result of further improvements in the performance of the confinement technique, squeezing (see 8.4.) between uncertainties of the momentum and the

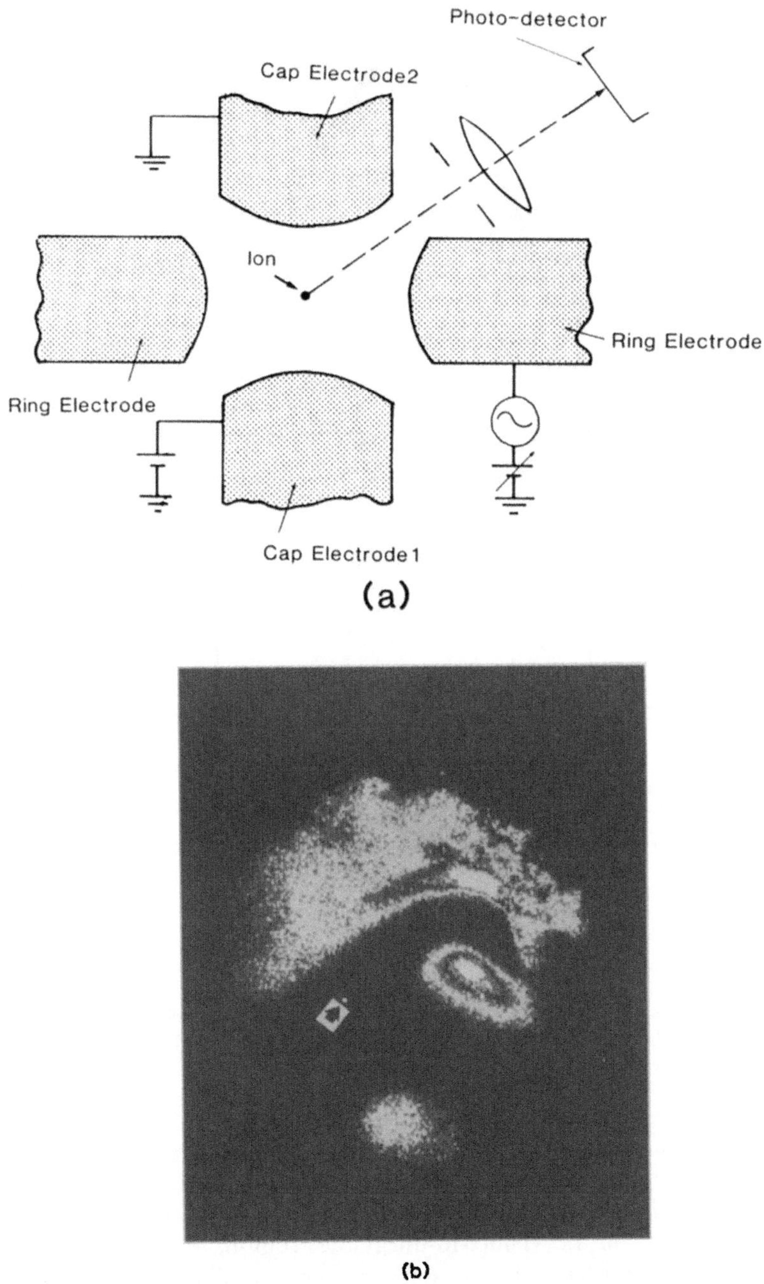

Fig. 7.28. (a) Experiment combining the laser cooling and the ion trap for stopping ions (after Itano, Bergquist, and Wineland [80]). (b) Hg^+ ion at rest (by courtesy of W. Itano, National Institute of Standard and Technology, 1990); arrow points to the fluorescence from a single ion.

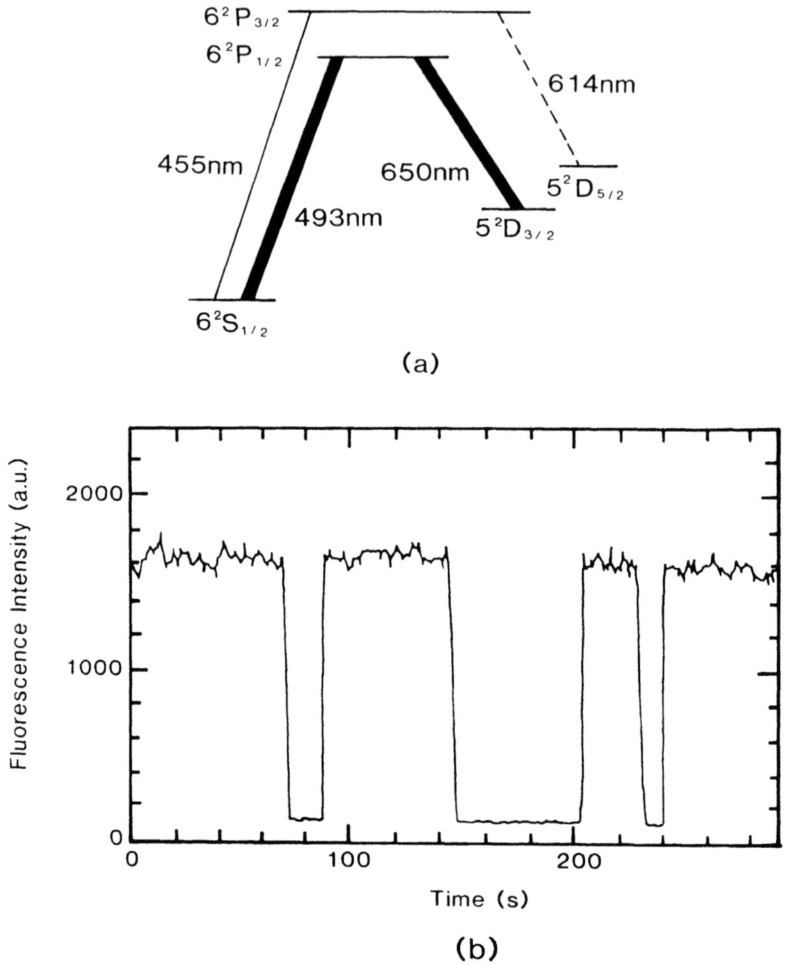

Fig. 7.29. Experimental result of the quantum jump (after Nagourney, Sandberg and Dehmelt [83]). (a) Energy levels of Ba ion. (b) Experimental result. The high and low levels of the fluorescent intensity correspond to the transitions from $6^2P_{1/2}$ to $6^2S_{1/2}$ and $5^2D_{1/2}$, respectively.

position of the ion was recently observed [85]. Furthermore, by confining and cooling high density atoms, it was observed that their characteristics were similar to those of ionic crystal and solid state plasma [86]. Since most ions have their resonant frequencies in the visible-to-ultraviolet region, several expensive dye lasers and their second harmonic or sum-frequency radiations have been used. However, the efficiency of these wavelength conversion is rather low in these short wavelength regions. To overcome these difficulties, the possibility of using easily controllable semiconductor lasers has been examined [85].

(4) Cavity quantum electrodynamics (Cavity QED): As was described in 5.4, the enhancement or inhibition of spontaneous emission from atoms can be achieved by using a micro-cavity with the size of wavelength dimension. At the initial stage of the cavity QED study, experiments were carried out in the microwave frequency region [87]. In these experiments, several AlGaAs lasers were used as multi-step optical pumping sources to excite the atoms to the Rydberg states so that they would have a large transition dipole moment. A controlled rate of spontaneous emission and specific Rabi oscillation driven by vacuum fluctuations were observed by using a micro-cavity and single atom maser [87]. To prepare the atoms with a large transition dipole moment, Cs atoms were excited to the Rydberg state by using AlGaAs lasers as pumping sources. Due to the technical difficulty to fabricate a micro-cavity in the optical frequency region, experiments of optical cavity QED were carried out by installing atoms in a high finesse and mode-degenerated confocal Fabry-Perot cavity. As shown in Fig. 7-30, the enhancement and inhibition of spontaneous emission from Yb atoms was observed at resonant and non-resonant frequencies of the confocal Fabry-Perot cavity, respectively [88].

The experimental study of single atom lasers has also been progressing steadily. As shown in Fig. 7-31, FM sidebands on the spontaneous emission spectral line from Na atoms in the high finesse Fabry-Perot cavity have recently been

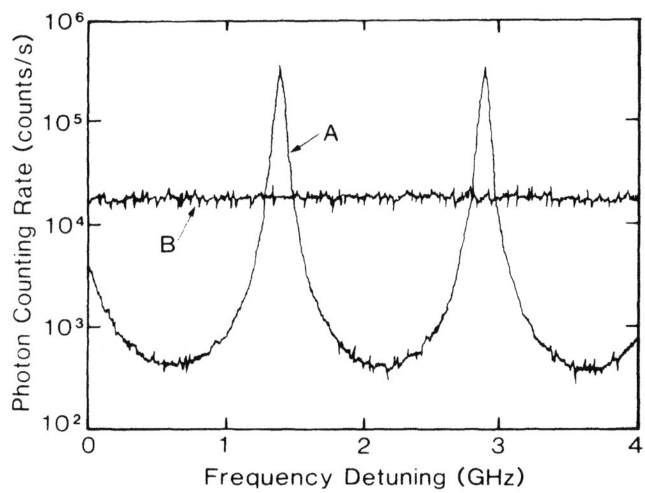

Fig. 7.30. Experimental result of the spontaneous emission control of Yb atoms in a Fabry-Perot resonator (after Heinzen, Childs, Thomas, and Feld [88]). The laser frequency irradiated into Yb atoms coincides with the resonance frequency of Yb atoms. In the figure, vertical axis corresponds to the spontaneous emission intensity of Yb atoms in a confocal Fabry-Perot resonator,and horizontal axis corresponds to the detuning between the resonance frequency of Yb atoms and that of Fabry-Perot resonator. Curve A shows the result obtained by using the Fabry-Perot cavity. Curve B shows the case that Fabry-Perot cavity is not used, i.e. the spontaneous emission intensity in free space.

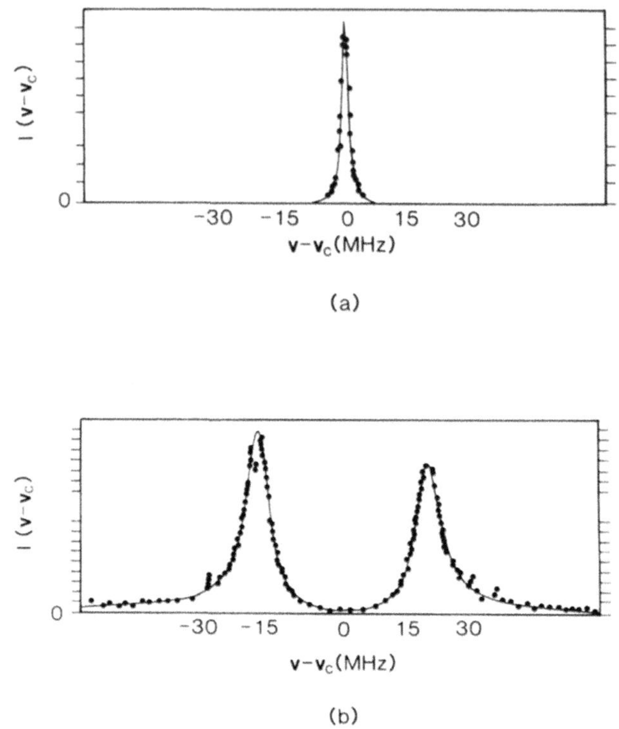

Fig. 7.31. Experimental results of measuring the vacuum Rabi oscillation in Na atoms (after Raizen, Thompson, Brecha, Kimble, and Carmichael [89]). (a) The spectral profile $I(\nu-\nu_c)$ of the empty Fabry-Perot cavity. (b) The spectral profile $I(\nu-\nu_c)$ of the cavity with Na atoms. This profile has two peaks representing the FM sidebands due to the vacuum Rabi oscillation. ν_c is the cavity resonance frequency.

observed and are attributed to vacuum Rabi oscillations [89]. (For further details, see 8.5.) The cavity QED technique could become useful in developing new light-emitting semiconductor devices, e.g. low threshold semiconductor lasers. For this purpose, the Fabry-Perot micro-cavity with optical wavelength dimension has been fabricated, as shown in Fig. 7-32 [90]. The enhancement and inhibition of spontaneous emission from a semiconductor quantum well (QW) have also been demonstrated by using multi-layer distributed Bragg reflectors as the a cavity mirrors [91]. Since the study of optical cavity QED is still in its the early stage of development, there are several problems remaining to be solved. For example, conventional studies on cavity QED have demonstrated that spontaneous emission is enhanced at the cavity resonance frequency. Although this enhancement may increase the quantum efficiency of the laser, it should be inhibited in order to reduce the quantum noise and to realize a low FM noise laser. It is expected that these problems could be solved within a near future and that new and stable semiconductor lasers could be developed by using the cavity QED technique.

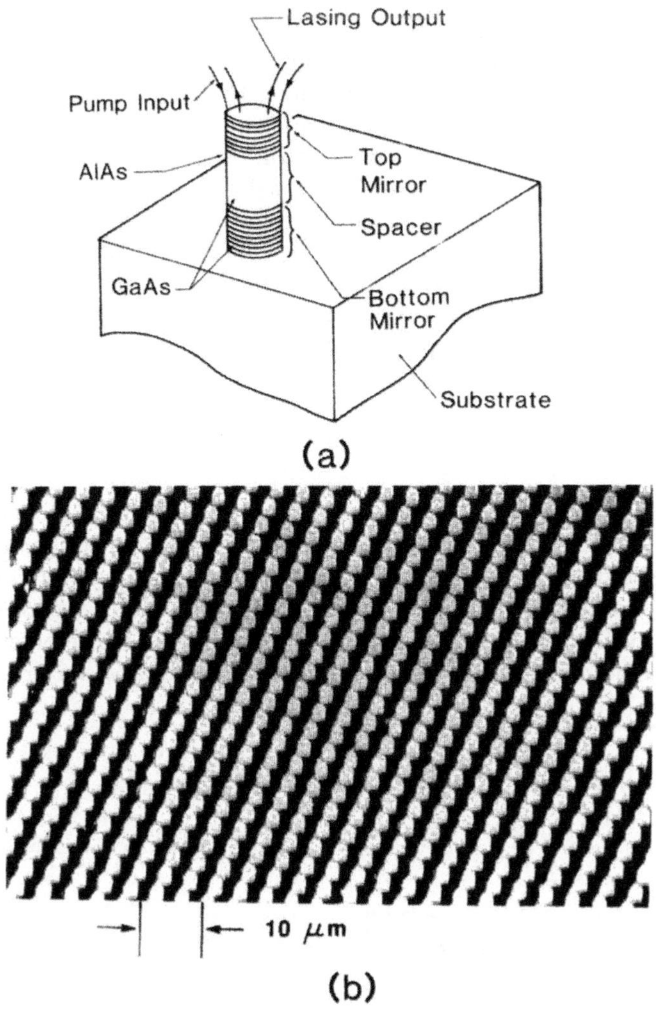

Fig. 7.32. A Fabry-Perot microcavity employing a semiconductor device fabrication technique. (a) Schematic of the device structure (after Jewell, McCall, Lee, Scherer, Gossard, and English [90]). Its diameter and length are 1.5 μm and 4.0 μm, microcavities. (b) Photograph of the two-dimensional array of the microcavities (by the courtesy of J. Jewell, AT & T Bell Laboratories, 1991).

(5) Detection of gravitational wave and test of general theory of relativity: Gravitational wave detection is an effective tool to test the general theory of relativity. A heavy metal rod (called a Weber bar) hung by metal wires has been used as the antenna of detection [92]. However, its detection bandwidth is narrow because the Weber bar works as a band-pass filter whose resonant frequency is determined by its size and weight. To expand the bandwidth, the method of using

a laser interferometer has been proposed. An example of this experimental setup is shown in Fig. 7-33, in which a long baseline interferometer has been constructed by using a low FM noise laser as a light source. The gravitational wave is generated from a neutron star or by the gravity collapse in the galaxy and propagates to the earth. If the gravitational wave hits one of the interferometer mirrors hung by a wire, the difference in optical path lengths between the two interferometer arms changes slightly. As a result, the output light power of the interferometer changes, by which the gravitational wave can be detected. Frequent measurements can be carried out if the amplitude of a gravitational wave d of 1×10^{-21} is measured because such a gravitational wave is generated about once a month by the gravity collapse in the Virgin galaxy. The relationship between the amplitude d and the displacement of the interferometer mirror δl is given by

$$d = 2\delta l / l \qquad (7-1)$$

where l is the length of the interferometer. This equation shows that the value of l should be 100 km since the values of d and the minimum displacement measured by the low FM noise laser are 1×10^{-21} and 1×10^{-14} (cm), respectively. Since it is technically difficult to construct such a long interferometer, an interferometer with a folded optical path or a high finesse Fabry-Perot interferometer has been proposed. Furthermore, a 10,000 km-long baseline interferometer has also been

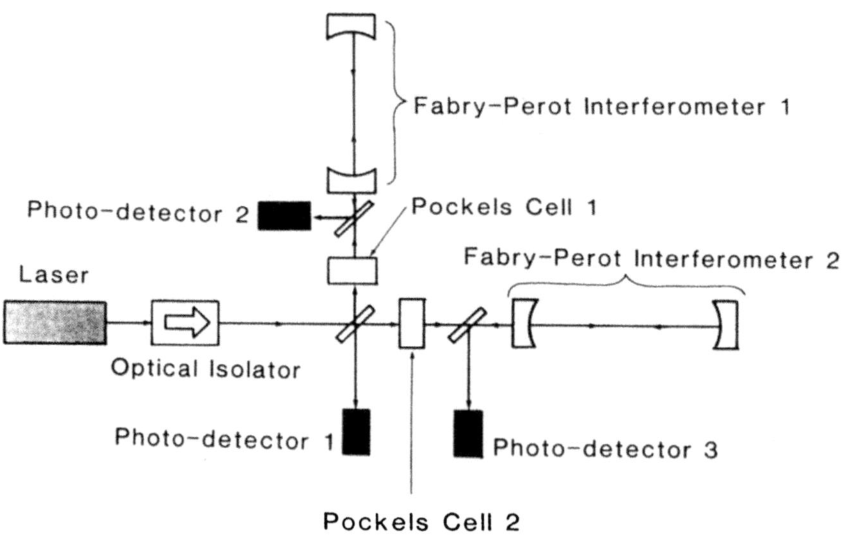

Fig. 7.33. Setup for measurement of the gravitational wave with very long Fabry-Perot interferometers. The gravitational wave induces a slight movement of the mirrors of the two Fabry-Perot interferometers. The variation of the interference fringe power is detected by the photo-detector 1.

proposed to be constructed in space by using satellites. A simultaneously indispensable consideration is that a laser with a frequency stability higher than 1×10^{15} is also required. Furthermore, as is represented by Eq.(1-71), a power higher than 100 W is required since the accuracy of phase measurement is proportional to light power. For such a light source satisfying these requirements, a semiconductor laser-pumped Nd : YAG laser, as described in 3.3, is considered to be a promising candidate. By using a low power and frequency-controlled Nd : YAG laser as a master laser, frequency tracking of the high power slave laser is carried out by using the technique described in 6.3.1(d) or by the technique of injection locking. Based on these ideas, large experimental systems for gravitational wave detection are under construction recently and success in detection is expected at the beginning of the 21st century. The appearance of a preliminary experimental setup, using a 10 m-long Michelson interferometers and an Ar^+ laser, is shown by Fig. 7-34.

Fig.7.34. View of a preliminary laser interferometer system for gravitational wave detection (by the courtesy of N. Kawashima, the Institute of Space and Astronomical Science, 1989).

For further improvements of the phase detection sensitivity, the use of squeezed light and quantum non-demolition measurement, as described in chapter 8, have been studied.

[Textbooks]

[A] M. Ohtsu, *Lasers and Atomic Clocks*, Ohm-sha Publishing, Tokyo, 1986 (in Japanese).

[References]

1) J. P. Gordon, "Quantum Effects in Communication Systems", *Proc. IRE*, Vol. 50 (1962) pp. 1898–1908.
2) B. S. Glance and O. Scaramucci, "High-Performance Dense FDM Coherent Optical Network", *IEEE J. on Selected Areas In Commun.*, Vol. 8 (1990) pp. 1043–1047.

3) M. Ohtsu, M. Murata, and M. Kourogi, "Subkilohertz linewidth of a semiconductor laser by electrical feedback and its network analysis", *Proc. of the Conference on Lasers and Electro-Optics* (*CLEO'90*), April 1989, Baltimore (Opt. Soc. Am., Washington, D.C.) THK30.

4) Y. -H. Cheng, T. Okoshi, and O. Ishida, "Performance Analysis and Experiment of a Homodyne Receiver Insensitive to Both Polarization and Phase Fluctuations", *J. Lightwave Technol.*, Vol. 7 (1989) pp. 368–374.

5) R. G. Marshalek and G. A.Koepf, "Comparison of optical technologies for intersatellite links in a global telecommunications network", *Opt. Eng.*, Vol. 27 (1988) pp. 663–676.

6) M. Ohtsu and S. Araki, "Using a 1.5 μm DFB InGaAsP laser in a passive ring cavity-type fiber gyroscope", *Appl. Opt.*, Vol. 26 (1987) pp. 464–470.

7) G. Sagnac, "L'ether luminux demontre par l'effect du vent relatif d'ether dams un interferometre en rotation uniforme", *Compt. Rend.*, Vol. 157 (1913) pp. 708–717.

8) S. Tai, M. Takahashi, and K. Kyuma, "Drift Characteristics of Fiber-Optic Passive Ring-Resonator Gyro", Reports on Topical Meeting of the Laser Soc. Jpn., Feb. 1989, Tokyo, (Laser Soc. Jpn., Osaka) RTM-88-40 (in Japanese).

9) C. -H. Shin, M. Teshima, M. Ohtsu, T. Imai, J. Yoshida, and K. Nishide, "Modulatable, High Coherent and Compact Semiconductor Laser Modules", *Proc. of the Seventh International Conference on Integrated Optics and Optical Fiber Communication* (IOOC'89), July 1989, Kobe, 21D4-5.

10) J. Yoshida, T. Imai, K. Nishide, C. -H. Shin, M. Teshima, and M. Ohtsu, "Modulatable High Coherent Laser for Fiber Optic Gyroscope", Reports on Topical Meeting of the Laser Society of Japan, Feb. 1989, Tokyo (Laser Soc. Jpn., Osaka) RTM-88-37 (in Japanese).

11) C. -H. Shin, M. Teshima, M. Ohtsu, T. Imai, J. Yoshida, and K. Nishide, "FM Characteristics and Compact Module for Coherent Semiconductor Lasers Coupled to an External Cavity", *IEEE Photonics Technol. Lett.*, Vol. 2 (1990) pp. 167–169.

12) T. Nishimura, "Micro-measurement systems by lasers", *Proc. of the Seminar on Optical Measurements*, May 1989, Tokyo (Soc. Optical Measurement Instruments of Jpn., Tokyo, 1989) pp. 1–10 (in Japanese).

13) G. Binnig and H. Rohrer, "Scanning tunneling microscopy", *Helv. Phys. Acta*, Vol. 55 (1982) pp. 726–735.

14) G. Binnig, C. F. Quate, and Ch. Gerber, "Atomic Force Microscope", *Phys. Rev. Lett.*, Vol. 56 (1986) pp. 930–933.

15) D. W. Pohl, W. Denk, and M. Lanz, "Optical Stethoscopy: Image recording with resolution λ/20", *Appl. Phys. Lett.*, Vol. 44 (1984) pp. 651–653.

16) R. C. Reddick, R. J. Warmack, and T. L. Ferrell, "New form of scanning optical microscopy", *Phys. Rev. B*, Vol. 39 (1989) pp. 767–770.

17) M. Ohtsu, K. Nakagawa, S. Jiang, and N. Tomita, "Super-Resolution Photon Scanning Tunneling Microscope", Technical Digest of the 3rd Optoelectronics Conference, July 1990, Makuhari Messe (Inst. Electron., Information and Commun. Eng., Tokyo, 1990) 12D1-1.

18) S. Jiang, N. Tomita, K. Nakagawa, and M. Ohtsu, "Proposal of Super-Sensitive Reflection-Mode Phase-Locked Photon Scanning Tunneling Microscope by Diode Lasers", Technical Digest of the 3rd Optoelectronics Conference, July 1990, Makuhari Messe (Inst. Electron., Information and Commun. Eng., Tokyo, 1990) 12D1-3.

19) M. Fukuda, I. Ogasawara, A. Nishimura, H. Suganuma, and S. Suzuki, "Characteristics of crosstalk in four-core graded-index multicore fiber", Technical Digest of Optical Fiber Communication Conference / The 6th International Conference on Integrated Optics and Optical Fiber Communication (OFC/IOOC'87), Jan. 1987, Reno (Opt. Soc. Am., and LEOS of IEEE, Washington, D.C., 1987) WI7.

20) N. Takeuchi, "Development of Pseudo-Random Modulation CW Lidar and Its Application to Field Measurements", Res. Rep. from the Nat. Inst. for Environmental Studies, Japan, No. 122, 1989 (in Japanese).

21) S. Okano, M. Taguchi, and H. Fukunishi, "Development of a Laser Heterodyne Spectrometer Using a Tunable Diode Laser as a Local Oscillator (III)", *Proc. The 13th Jpn. Laser Sensing Symposium, Oct. 1989, Fukuoka* (Laser Radar Soc. Jpn., Tokyo, 1989) B7 (in Japanese).
22) C. Yamanaka (Chief Editor), "*Laser Handbook*", Ohm-sha Publishing, Tokyo 1978 (in Japanese).
23) A. D. Streatere, J. Mooibroek, and J. P. Woerdman, "Enhancement efficiency in separation of Rb isotope by light-induced drift with the use of a diode laser with relaxation sidebands", *Appl. Phy. Lett.*, Vol. 52 (1988) pp. 602–604.
24) J. Lawrenz, A. Obrebski, and K. Niemax, "Measurement of Isotope Ratios by Doppler-Free Laser Spectroscopy, Applying Semiconductor Diode Lasers and Thermoionic Diode Detection", *Anal. Chem.*, Vol. 59 (1987) pp. 1232–1236.
25) M. Ohtsu, "Investigation on Applicability of Visible Semiconductor Lasers to Elementary Analysis", *Proc. the 1st Meeting on Lightwave Sensing Technol., June 1988, Tokyo* (Lightwave Sensing and Technol. Res. Soc. Group, Jpn. Soc. Appl. Phys., Tokyo) LST1-4 (in Japanese).
26) M. Ohtsu, H. Suzuki, K. Nemoto, and Y. Teramachi, "Narrow-Linewidth Tunable Visible InGaAlP Laser, Application to Spectral Measurements of Lithium, and Power Amplification", *Jpn. J. Appl. Phys.*, Vol. 29 (1990) pp. L1463–L1465.
27) G. Inoue and M. Suzuki, "Laser-induced fluorescence of the SiH_2 radical", *Chem. Phys. Lett.*, Vol. 105, No. 6 (1984) pp. 641–644.
28) J. L. Picque, "Hyperfine optical pumping of a cesium atomic beam, and applications", *Metrologia*, Vol. 13 (1977) pp. 115–119.
29) S. Ohsima, Y. Nakadan, and Y. Koga, "The optically pmmped Cs frequency standard at the NRLM", *IEEE Trans. Instrum. Meas.*, Vol. IM-37 (1988) pp. 409–413.
30) N. F. Ramsey, "*Molecular Beams*", (Oxford Univ. Press, Oxford, 1956).
31) Y. Nakadan, S. Ohshima, T. Ikegami, and Y. Koga, "Characteristics of the optically pumped cesium beam frequency standard at NRLM (I)", Extended Abstracts of the 49th Autumn Meeting, Oct. 1988, Toyama (Jpn. Soc. Appl. Phys., Tokyo, 1988) 4aZL1 (in Japanese).
32) By the courtesy of Y. Koga, National Res. Lab. of Metrology, 1989.
33) D. W. Sesko and C. E. Wieman, "Observation of the cesium clock transition in laser-cooled atoms", *Opt. Lett.*, Vol. 14 (1989) pp. 269–271.
34) M. Hashimoto and M. Ohtsu, "Experiments on a semiconductor laser pumped rubidium atomic clock", *IEEE J. Quantum Electron.*, Vol. QE-23 (1987) pp. 446–451.
35) M. Hashimoto and M. Ohtsu, "Modulation transfer and optical Stark effect in a rubidium atomic clock pumped by a semiconductor laser", *J. Opt. Soc. Am. B*, Vol. 6 (1989) pp. 1777–1789.
36) M. Ohtsu, M. Hashimoto, and H. Suzuki, "Performance evaluation of a semiconductor laser pumped Rb atomic clock", Extended Abstract of the 50th Autumn Meeting of the Jpn. Soc. Appl. Phys., 1989 (Jpn. Soc. Appl. Phys., Tokyo, 1989) 29aX5 (in Japanese).
37) M. Hashimoto and M. Ohtsu", A novel method to compensate for the effect of light shift in a rubidium atomic clock pumped by a semiconductor laser", *IEEE Trans. Instrum. Meas.*, Vol. 39 (1990) pp. 458–462.
38) M. Ohtsu, M. Hashimoto, H. Furuta, and H. Suzuki, "Performance evaluations of a semiconductor laser pumped rubidium atomic clock", The Second Workshop on the Precise Phase Locked Network June 1989, Tokyo (Institute of Electric Eng. Jpn., Tokyo, 1989) paper 2-1 (in Japanese).
39) H. Furuta, H. Suzuki, and M. Ohtsu, "Observation of Ramsey-Type Resonant Fringe Using a Cylindrical Microwave Cavity for a Diode Laser-Pumped Rb Beam Atmic clock", *Jpn. J. Appl. Phys.*, Vol. 30 (1991), pp. 596–602.
40) W. Lenth, "Optical heterodyne spectroscopy with frequency- and amplitude-modulation semiconductor lasers", *Opt. Lett.*, Vol. 8 (1983) pp. 575–577.
41) M. Ohtsu, H. Kotani, and H. Tagawa, "Spectral Measurements of NH_3 and H_2O for Pollutant Gas Monitoring by 1.5 μm InGaAsP/InP Lasers", *Jpn. J. Appl. Phys.*, Vol. 22 (1983) pp. 1553–1557.
42) K. Fukuoka, M. Ohtsu, and T. Tako, "Accurate Wavelength Measurement of the Absorption Lines in H_2O Vapor by a 0.8 μm AlGaAs Laser", *Jpn. J. Appl. Phys.*, Vol. 23 (1984) pp. L117–L120.

43) B. T. Darling and D. M. Dennison, "The water vapor molecule", *Phys. Rev.*, Vol. 57 (1940) pp. 128–139.

44) S. Kinugawa, H. Sasada, and K. Uehara, "Detection of C_2H_2 absorption lines with 1.5 μm DFB lasers", Extended Abstracts of The 49th Autumun Meeting, Sept. 1988 (The Jpn. Soc. Appl. Phys.) 6pQ12 (in Japanese).

45) H. Sasada, "1.5 μm DFB semiconductor laser spectroscopy of HCN", *J. Chem. Phys.*, Vol. 88 (1988) pp. 767–777.

46) K. Nakagawa, "Localization of a second harmonic vibration in a NH_3 molecule", Ph.D. Thesis, 1989, Univ. of Tokyo.

47) G. C. Bjorklund, "Frequency-modulation spectroscopy: a new meethod for measuring weak absorption and dispersion", *Opt. Lett.*, Vol. 5 (1980) pp. 15–17.

48) L. -G. Wang, H. Riris, C. B. Carlisle, and T. F. Gallangher, "Comparison of approaches to modulation spectroscopy with GaAlAs semiconductor lasers: application to water vapor", *Appl. Opt.*, Vol. 27 (1988) pp. 2071–2077.

49) H. Tsuchida, M. Ohtsu, Tako, N. Kuramochi, and N. Oura, "Frequency stabilization of AlGaAs semiconductor laser based on the ^{85}Rb-D_2 line", *Jpn. J. Appl. Phys.*, Vol. 21 (1982) pp. L561–L563.

50) T. Yabuzaki, A. Ibaraki, H. Hori, M. Kitano, and T. Ogawa, "Frequency-locking of a GaAlAs laser to a Doppler-free spectrum of the Cs-D_2 line", *Jpn. J. Appl. Phys.*, Vol. 20 (1981) pp. L451–L454.

51) S. Yamaguchi and M. Suzuki, "Frequency stabilization of a diode laser by use of the optogalvanic effect", *Appl. Phys. Lett.*, Vol. 41 (1982) pp. 597–598.

52) Y. C. Chung, "Frequency-locking of a 1.3 μm DFB laser using a miniature argon glow lamp", *IEEE Photonics Technol. Lett.*, Vol. 1 (1989) pp. 135–136.

53) K. Sakurai and N. Yamada, "Second-harmonic radiation from an AlGaAs laser and its application for absorption spectroscopy", *Opt. Lett.*, Vol. 14 (1989) pp. 233–235.

54) M. Ohtsu and E. Ikegami, "Frequency stabilization of 1.5 μm DFB laser using internal second harmonic generation and atomic ^{87}Rb line", *Electron. Lett.*, Vol. 25 (1989) pp. 22–23.

55) H, Suzuki, I. Koshiishi, M. Ohtsu, and Y. Teramachi, "Frequency Control of Visible Semiconductor Laser", Extended Abstracts of the 50th Autumun Meeting, Sept. 1989 (The Jpn. Soc. Appl. Phys., Tokyo) 29pZL7 (in Japanese).

56) M. Ohtsu, M. Hashimoto, and H. Ozawa, "A highly stabilized semiconductor laser and its application to optically pumped Rb atomic clock", *Proc. the 39th Annual Symposium of Frequency Control, Philadelphia, 1985* (IEEE, Piscataway, NJ, 1985) pp. 43–53.

57) M. Suzuki and S. Yamaguchi, "Frequency stabilization of a GaAs semiconductor laser by use of the optical-optical double resonance effect of the Doppler-free spectrum of the Rb D_1 line", *IEEE J. Quantum Electron.*, Vol. 24 (1988) pp. 2392–2399.

58) E. Ikegami, H. Kusuzawa, H. Furuta, and M. Ohtsu, "Frequency link between InGaAsP and GaAs lasers by using the Doppler-free optical-optical double resonance in Rb", Extended Abstracts of the 37th Spring Meeting of the Jpn. Soc. Appl. Phys., March 1990 (The Jpn. Soc. Appl. Phys., Tokyo, 1990) 31aG3 (in Japanese).

59) C. Vadla, A. Obrebski, and K. Niemax, "Isotope shift of the $3s^2S_{1/2}$ and $3p^2S_J$ levels in ^{6,7}Li", *Opt. Commun.*, Vol. 63 (1987) pp. 288–292.

60) M. Ohtsu and T. Tako, "Coherence in semiconductor lasers", in *Progress in Optics XXV*, Ed. by E. Wolf, 1988 (Elsevier, Amsterdam, 1988) pp. 191–278.

61) H. Rinnenberg, "Ultrahigh n Rydberg atoms", Technical Digest of the 14th Int. Conf. on Quantum Electronics, San Francisco, 1986 (LEOS of IEEE, OSA, Washington, D.C., 1986) THKK1.

62) S. L. Gilbert and C. E. Wieman, "Atomic-beam measurement of parity nonconservation in cesium", *Phys. Rev. A*, Vol. 34 (1986) pp. 792–803.

63) M. C. Noecker, B. P. Masterson, and C. E. Wieman, "Precision measurement of parity nonconservation in atomic cesium: a low-energy test of the electroweak theory", *Phys. Rev. Lett.*, Vol. 61 (1988) pp. 310–313.

64) T. P. Emmons, J. M. Reeves, and E. N. Fortson, "Parity-nonconserving optical relation in atomic lead", *Phys. Rev. Lett.*, Vol. 51 (1983) pp. 2089–2092.
65) C. E. Tanner and C. E. Wieman, "Precision measurement of the Stark shift in the $6S_{1/2} \rightarrow 6P_{3/2}$ cesium transition using a frequency-stabilized laser diode", *Phys. Rev. A*, Vol. 38 (1988) pp. 162–165.
66) S. Chu, AT & T Bell Laboratories, private communication (1986).
67) T. W. Hänsch and A. L. Schawlow, "Cooling of gases by laser radiation", *Opt. Commun.*, Vol. 13 (1975) pp. 68–69.
68) P. D. Lett, R. N. Watts, C. I. Westbrook, W. D. Phillips, P. L. Gould, and H. J. Metcalf, "Observation of atoms laser cooled below the Doppler limit", *Phys. Rev. Lett.*, Vol. 61 (1988) pp. 169–172.
69) R. N. Watts, P. D. Lett, C. I. Westbrook, S. L. Rolston, C. E. Tanner, and W. D. Phillips, "Laser cooling below the Doppler limit", *Proc. Conference on Quantum Electronics and Laser Science (QELS'89)*, Baltimore, 1989 (Opt. Soc. Am., Washington, D.C., 1989) THLL1.
70) A. Aspect, E. Arimond, R. Kaiser, N. Vansteenkiste, and C. Cohen-Tannoudji, "Laser cooling below the one-photon recoil energy by velocity-selective coherent population trapping", *Phys. Rev. Lett.*, Vol. 61 (1988), pp. 826–829.
71) K. Shimoda, "Atoms in Precision Electromagnetic Measurements", *IEEE Trans. Instrum. Meas.*, Vol. IM-38 (1989) pp. 150–155.
72) M. A. Kasevich, D. S. Weiss, and S. Chu, "Normal-incidence reflection of slow atoms from an optical evanescent wave", *Opt. Lett.*, Vol. 15, No. 11 (1990) pp. 607–609.
73) E. L. Raab, M. G. Prentiss, A. E. Cable, S. Chu, and D. E. Pritchard, "Trapping of neutral sodium atoms with radiation pressure", *Phys. Rev. Lett.*, Vol. 59 (1987) pp. 2631–2634.
74) D. Sesko, C. G. Fan, and C. E. Wieman, "Production of a cold atomic vapor using diode-laser cooling", *J. Opt. Soc. Am. B*, Vol. 5 (1988) pp. 1225–1227.
75) R. N. Watts, D. H. Yang, B. Sheehy, and H. Metcaff, "Deceleration and cooling of a thermal beam of rubidium", *Proc. the 15th International Quantum Electronics Conference (IQEC'87)* Baltimore, 1987 (Opt. Soc. Am., Washington, D.C., 1987) TUGG37.
76) H. Katori, T. Yamashita, and F. Shimizu, "Laser cooling of inert gases by semiconductor lasers", *Proc. Annual Meeting of Jpn. Soc. Phys., October 1989* (Jpn. Soc. Phys., Tokyo, 1989) 5aZF11 (in Japanese).
77) C. E. Wieman, Joint Institute for Laboratory Astrophysics, Unov. Colorado, private communication (1989).
78) R. G. Beausoleil and T. W. Hänsch, "Two-photon optical Ramsey spectroscopy of freely falling atoms", *Opt. Lett.*, Vol. 10 (1985) pp. 547–549.
79) M. A. Kasevich, E. Riis, S. Chu, and R. G. Devoe, "rf Spectrocopy in an Atomic Fountain", *Phys. Rev. Lett.*, Vol. 63 (1989) pp. 612–614.
80) W. M. Itano, J. C. Bergquist, and D. J. Wineland, "Laser Spectroscopy of Trapped Atomic Ions", *Science*, Vol. 237 (1987) pp. 612–617.
81) E. Schrödinger, "Are there quantum jumps?, Part I", *British J. Phil. Sci.* Vol. 3 (1952) pp. 109–123.
82) J. C. Bergquist, R. G. Hulet, W. M. Itano, and D. J. Wineland, "Observation of Quantum Jump in a Single Atom", *Phys. Rev. Lett.*, Vol. 57 (1986) pp. 1699–1702.
83) W. Nagourney, J. Sandberg, and H. Dehmelt, "Shelved Optical Electron Amplifier: Observation of Quantum Jumps", *Phys. Rev. Lett.*, Vol. 56 (1986) pp. 2797–2799.
84) Th. Sauter, W. Neuhauser, R. Blatt, and P. E. Toschek, "Observation of Quantum Jumps", *Phys. Rev. Lett.*, Vol. 57 (1986) pp. 1696–1698.
85) J. Bergquist, National Institute of Standards and Technol., 1989, private communication.
86) S. L. Gilbert, J. J. Bollinger, and D. Wineland, "Self-Avoiding Walks on Randomly Diluted Lattices", *Phys. Rev. Lett.*, Vol. 60 (1988) pp. 2022–2025.
87) D. Meschede, H. Walther, and G. Muller, "One-Atom Maser", *Phys. Rev. Lett.*, Vol. 54 (1985) pp. 551–554.

88) D. J. Heinzen, J. J. Childs, J. F. Thomas, and M. S. Feld, "Enhanced and Inhibited Visible Spontaneous Emission by Atoms in a Confocal Resonator", *Phys. Rev. Lett.*, Vol. 57 (1987) pp. 1320–1323.

89) M. Raizen, R. J. Thompson, R. J. Brecha, H. J. Kimble, and H. J. Carmichael, "Normal-mode splitting and linewidth averaging for two-state atoms in an optical cavity", *Phys. Rev. Lett.*, 63 (1989) pp. 240–243.

90) J. L. Jewell, S. L. McCall, Y. H. Lee, A. Scherer, A. C. Gossard, and J. H. English, "Lasing characteristics of GaAs microresonators", *Appl. Phys. Lett.*, Vol. 54 (1989) pp. 1400–1402.

91) K. Igeta, Y. Yamamoto, and S. Machida, "Control of spontaneous emission from a GaAs quantum well", Extended Abstracts of the Autumn Meeting of th Jpn. Soc. Phys., 1989 (Jpn. Soc. Phys., Tokyo, 1989) 4a-ZF-6 (in Japanese).

92) J. Weber, "Evidence for discovery of gravitational radiation", *Phys. Rev. Lett.*, Vol. 22 (1969) pp. 1320–1324.

Chapter 8

Toward the Future

Previous chapters described how to generate low FM noise lights and how to use them in several application systems. These lights correspond to lightwaves which are similar to classical sinusoidal waves. However, it is also possible to improve these performances further or to generate new lightwaves with specific performances different from those of these lightwaves. The recent progress in these new trials is reviewed and the problems remaining to be solved are presented in this chapter.

8.1 Expansion of the Frequency Range of Laser Oscillation

8.1.1 Short wavelength lasers

The quality factor for a laser can be defined as the ratio between the half linewidth $\Delta\nu_l$ and the center angular frequency ν_l of the field spectrum of a laser oscillation, which is given by

$$Q = \nu_l / \Delta\nu_l \qquad (8-1)$$

as was Eq.(2-16). Equation (4-50) shows that the linewidth $\Delta\nu_l (= 2\pi\Delta f)$ is inversely proportional to the oscillation power P, where P is proportional to the product of photon energy $\hbar\,\nu_l$ and photon number n. The photon number n in a stationary steady-state laser oscillation is given by Eq.(A-14) as a solution of the rate equation, where it can be easily found from Eq.(2-30) that the coefficient B_0 is independent of the optical frequency. Therefore, the value of Q is expressed by substituting Eq.(4-50) into Eq.(8-1) as

$$Q \propto \nu_l \qquad (8-2)$$

This equation shows that Q is larger for a higher optical frequency, i.e. the performance of the laser is higher for a higher frequency oscillation, from which it can be noted that the development of short wavelength laser is important.

However, Eq.(2-40) shows that the CW oscillation of a shorter wavelength is more difficult. To overcome this difficulty, wavelength conversion techniques employing nonlinear optical media have been used. The wavelength conversion efficiency is not sufficiently high when low FM noise CW lasers are used as primary light sources because their output power is not always sufficiently high. Therefore, as shown in Fig. 8-1, a nonlinear crystal such as $KNbO_3$ is installed in a high finesse Fabry-Perot cavity to build up the power of the primary laser light [1]. By using a 12 mW semiconductor laser with a 865 nm wavelength, a second harmonic wave of 0.2 mW power is obtained and corresponds to a conversion efficiency as high as 1.7%. The reflected light from this built-up cavity was injected into the laser to reduce the laser linewidth, which was based on the principle of optical feedback in Fig. 6-13. Figures 8-2(a) and (b) show an experimental setup by using a 100 mW AlGaAs laser with the 862 nm wavelength [2], which is similar to that of Fig. 8-1. A second harmonic wave of 3.3 mW is obtained, corresponding to the conversion efficiency of 3.3%. Furthermore, by employing the negative electrical feedback technique to this optically fedback laser, long-term stability of the second harmonic power is improved, as shown by Fig. 8-2(c). The rf voltage is applied to the $KNbO_3$ crystal so that an intensity-modulation efficiency of the second harmonic wave as high as 78% is obtained.

A MgO : $LiNbO_3$ crystal was polished and high-reflection films were coated on the end surfaces of the crystal to form a monolithic ring cavity, as is shown in Fig. 8-3. A second harmonic wave of 30 mW power was generated from this crystal by irradiating a semiconductor laser-pumped YAG laser of 53 mW power and this gave a wavelength conversion efficiency as high as 60% [3]. As an alternative candidate for high efficiency wavelength conversion, organic nonlinear optical crystals have been recently developed. Examples of these are listed in Table 8-1 [4]. These crystals are very promising, especially for semiconductor lasers whose

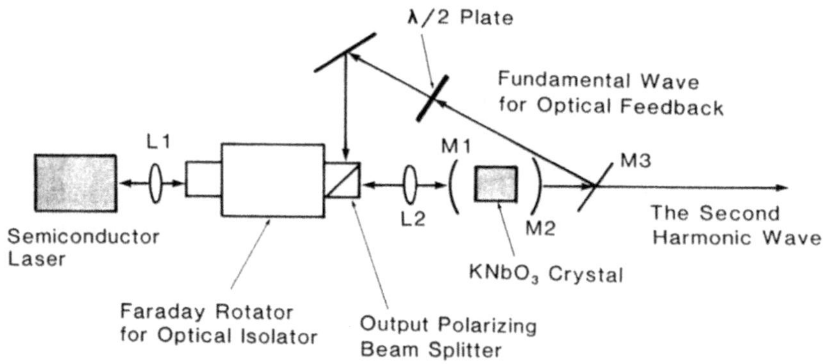

Fig. 8.1. Setup of the second harmonic generation of a semiconductor laser with a built-up cavity (after Dixon, Tanner, and Wieman [1]).

Name	Structure	Second Harmonic Intensity in the Case of Powder (Normalized to the Value of Carbon)	Molecular Polarization of the Second order ($\times 10^{-30}$ esu)
Urea	H_2NCONH_2	1.0	0.45
P-NA	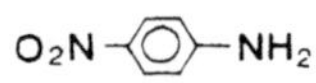	0.0	35
m-NA	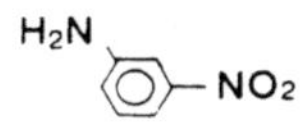	–	6
M-NA	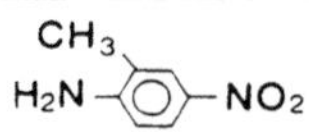	22	42
MAP	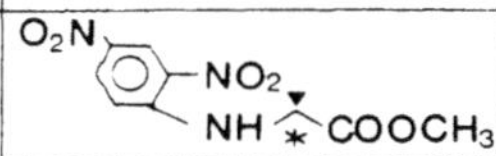	10	220
POM	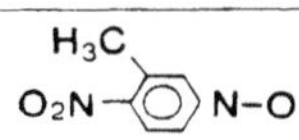	13	5
DAN	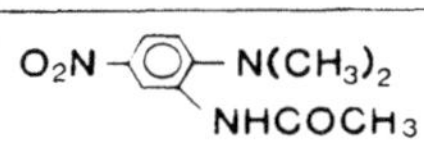	80	–
DANS	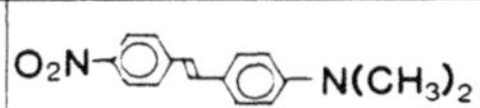	0	450
Merocyanine	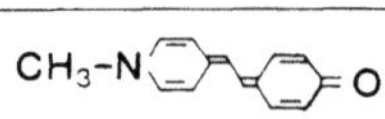	0	1000

Table 8.1 . Examples of organic nonlinear optical materials for high efficiency wavelength conversion (after Kajikawa and Takezoe [4]).

powers are generally lower than those of solid state lasers. The fabrication of nonlinear optical waveguides was tried by using these organic materials. It is advantageous to use these organic materials because a molecular design is possible to achieve the required performance. However, problems of reliability and lifetime still remain to be solved in the stage of fabrication of nonlinear optical devices such as waveguides. If these problems are solved in the future, it can be expected that highly efficient wavelength conversions will be possible.

8.1.2 Optical sweep generator

Wideband frequency sweeps have been rather difficult to obtain in conventional frequency stabilized lasers or even in microwave oscillators. Although an accurate and stable frequency sweep or tuning is possible by using the technique described in 6.3.1(e), its tuning range cannot be sufficiently wide. A stable frequency and accurate and wideband tunable lasers are the important light sources for a great number of application systems. The measure of the frequency tunable

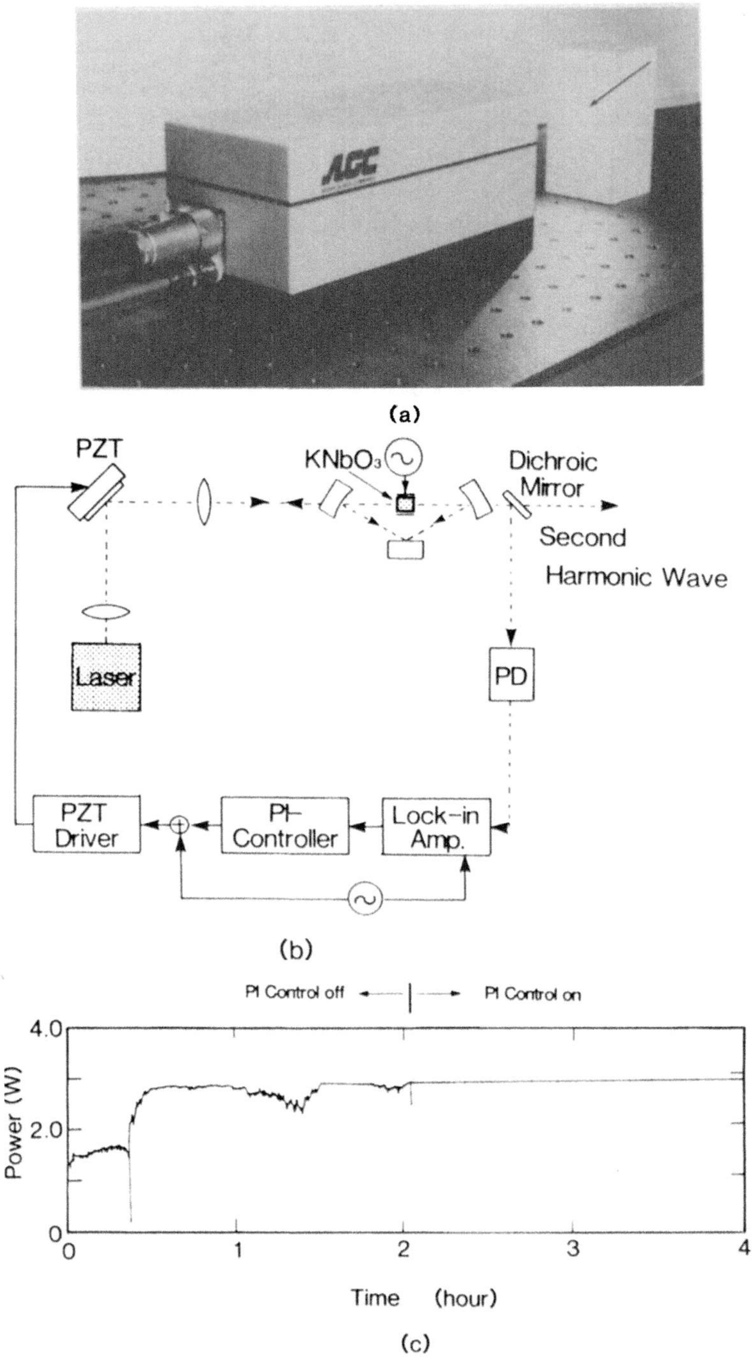
AGC
(a)
PZT
KNbO3
Dichroic
Mirror
Second
Harmonic Wave
Laser
PD
PZT
Driver
PI-
Controller
Lock-in
Amp.
(b)
PI Control off
PI Control on
4.0
Power (W)
2.0
0
0
1
2
3
4
Time (hour)
(c)

Fig. 8.2.

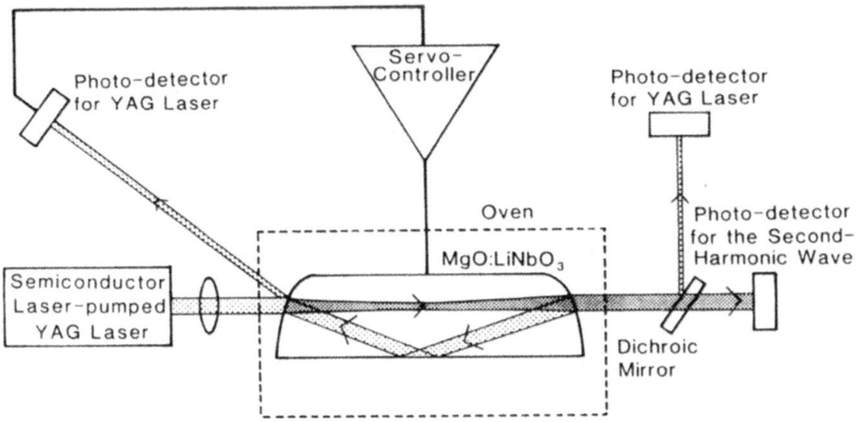

Fig. 8.3. Setup of the monolithic ring cavity for SHG of a semiconductor laser pumped YAG laser (after Kozlovsky, Nabors, and Byer [3], ©1988 IEEE).

range can be defined as

$$S_w = 2(\nu_u - \nu_l)/(\nu_u + \nu_l) \tag{8-3}$$

where ν_u and ν_l represent the upper and lower limits of tunable angular frequencies, respectively. The dye laser, conventionally used as a wavelength tunable laser, takes $S_w = 0.10$ if a single dye solution is used. For the semiconductor laser, the structure in Fig. 3-26(b) has obtained $S_w = 0.004$. Furthermore, by introducing a quantum well structure to the semiconductor laser device and by using an optical feedback technique using an external diffraction grating as a reflector, a wavelength tuning range over 100 nm has been realized [5]. This range gives a value of S_w as large as 0.12, which is comparable to that obtained by the dye laser.

Although the essential point is that a wideband tuning and high accuracy should be simultaneously realized, the examples demonstrated above do not necessarily satisfy this point. Figure 8-4 shows a schematic explanation of the system which has been proposed to realize such a high quality optical sweep generator [6]. Four kinds of commercially available lasers (with the wavelengths of 1.56 μm, 1.34 μm, 0.78 μm and 0.67 μm) are used as primary light sources and their FM noise reduction and frequency tracking are carried out by using the techniques

Fig. 8.2. Second harmonic generation of a 862 nm wavelength AlGaAs laser (after Senoh, Fujino, Tanabe, Ohtsu, and Nakagawa [2]): (a) appearance of the system (bright spot on the brick represents a spot of the second harmonic beam); (b) experimental setup; (c) measured temporal variation of the second harmonic power.

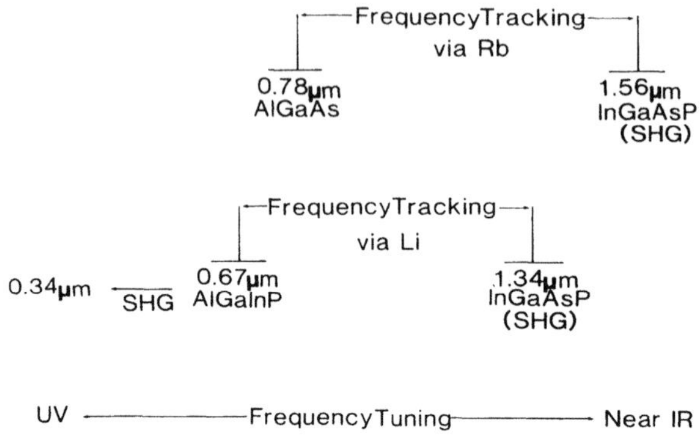

Fig. 8.4. Schematic explanation of a high precision and wide range optical sweep generator using four semiconductor lasers, Rb and Li atomic vapors, and nonlinear optical frequency conversion. The sweeping range is from 1.56 μm to 0.34 μm (after Ohtsu [6]).

described in 6.3.1. To realize a frequency link between two different wavelength lasers, a second harmonic wave generation is employed by using organic nonlinear optical waveguides. By using the optical-optical double resonance of Rb or Li atomic vapors, these laser frequencies are simultaneously locked to the common center frequency of the double resonance spectral lines (see Fig. 7-22). As a result, an optical frequency link is achieved. Coarse frequency tuning is realized by the techniques of sum- or difference- frequency generation and frequency conversions by parametric oscillation, where organic nonlinear optical waveguides are used. Fine tunings can be realized by the technique described in 6.3.1(e). By combining these techniques, it is estimated that tuning from the infrared (1.56 μm wavelength) to the ultra-violet (0.34 μm) is possible, which can be used to obtain a value of $S_{w'}$ as large as 1.28. Although an accurate optical sweep generator with a value of $S_{w'}$ larger than unity has not been developed so far, a very high quality system for a Peta-Hz (10^{15} Hz) class optical sweep generator can be obtained by using the controllability of semiconductor lasers, techniques of negative electrical feedback, nonlinear spectroscopy and nonlinear optical waveguides. Frequency link between 1.56 μm and 0.78 μm wavelength lasers, frequency control of a 0.67 μm wavelength laser and second harmonic generation have already been carried out [7].

8.2 Ultra-fast Detection of Lightwaves and Waveform Conversion

Since the optical frequency is much higher than that of a microwave, it is sometimes complicated to treat lightwaves. Although a fast optical frequency counter has not yet been realized, an optical frequency counting scheme has been employed by using the techniques of frequency division and heterodyne detection by using fast nonlinear photo-detectors, and of frequency comparison with a microwave oscillator such as a Cs atomic clock described in 7.4(1). The optical frequency of a 576 nm wavelength laser (521 THz) has been measured within an error of 1.6×10^{-10} by using a metal-insulator-metal (MIM) point contact diode composed of a tungsten wire and Nickel plate as a harmonic mixer and by using a nonlinear optical crystal as a second harmonic generator [8]. Although the response frequency of the MIM diode is as large as 200 THz (1.5 μm wavelength), the sensitivity is rather low (minimum detectable laser power is about 200 mW) and its lifetime is short. On the other hand, the response frequency of a Josephson device made by a superconductor-insulator-superconductor junction (SIS) is slow (4 THz) even though its sensitivity is high (its minimum detectable power is less than 1 mW) [9].

To overcome these difficulties in conventional photo-detectors, there are three candidates which may be used in the future. They are: (1) Using the Josephson junction of an oxide superconductor film [10], (2) Using the fast response of virtual polarization of the carriers in a quantum well semiconductor [11]; (3) Using the large nonlinear characteristics of an organic thin film [12]. Although a high sensitivity can be expected from (1) due to the large optical absorption coefficient of the oxide superconductor film, the problems remaining to be solved are the fabrication of a thin and homogeneous Josephson junction, and the development of a detailed theoretical model to analyze the dynamic response to a lightwave. In (2), the problems are the realization of a precise structure of multi-dimensional quantum confinement for the carrier and the improvement of the sensitivity of the dynamic response to the lightwave. In (3), the problems are the single crystal growth of an optimally designed organic material, the device fabrication and the improvement of lifetime. However, it can be expected that an ultra-high speed optical detector will be developed in the future by improving the theoretical background of oxide superconductors, sub-micron fine device fabrication techniques and so on. The full utilization of a high quality lightwave can be expected by using the Peta-Hz class optical sweep generator of 8.1.2 and these ultra-fast detectors.

A new scheme of optical frequency chain has been proposed recently for precise optical frequency measurements [13]. This scheme can be realized by using several semiconductor lasers and nonlinear optical crystals and will be used for the precise measurement of the Rydberg constant. Although this scheme does not require any ultra-fast photodetectors, a number of frequency-controlled semiconductor lasers and highly efficient nonlinear optical crystals are indispensable.

8.3 Generation of Non-Classical Photons and Their Applications

The characteristics of coherent state and hyper-coherent state have been described and their application systems have been demonstrated in the previous chapters. These lightwaves show quite low FM noises, as low as the quantum noise limit of each system, and are close to classical waves. In contrast to these, it is also possible to generate new lightwaves (non-classical photons) and to apply them to advanced application systems. They are described in this section.

8.3.1 Photon anti-bunching and squeezed state

When a constant intensity light beam is received by a photo-detector, the frequency of the photon arriving to the detector, i.e. the statistical properties of the time series of this arrival process, depends on the quantum state of light. As shown in Fig. 8-5 [14], the property of the coherent state of light is expressed as the Poisson

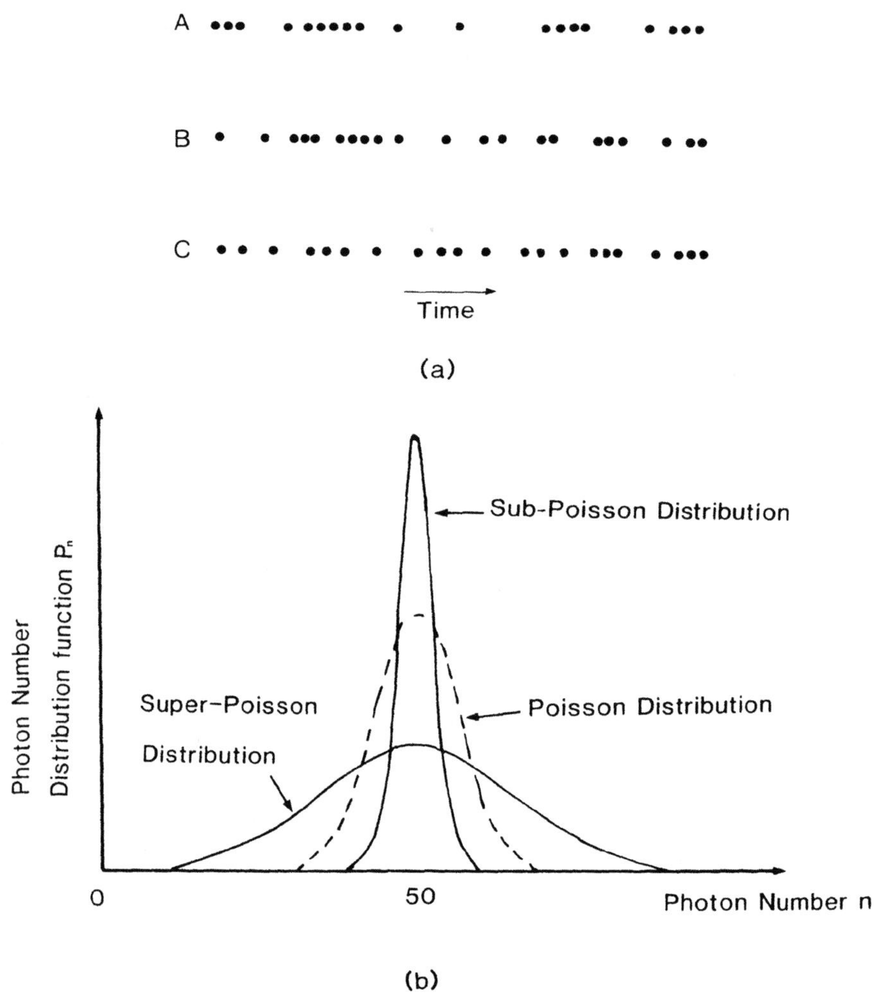

Fig. 8.5. Relation between photon number distribution and the temporal variation of photon counting (after Yajima [14]). (a) Temporal variation of photon counting representing the state of photon ensemble. The temporal variation of photo-electrons generated from the photon counting detector are shown by the dots. A: Bunching state. B: Random state. C: Anti-bunching state. (b) Distribution functions of photon numbers.

process because the photons arrive in a completely random manner while that of the incoherent thermal radiation is expressed as a Boltzmann distribution, i.e. photons arrive in a bunched manner. This phenomenon is called photon bunching. This is specific characteristic of classical light, which is represented as a statistical superposition of the coherent state of light. On the other hand, the phenomenon in which the photons do not bunch is called photon anti-bunching. This is the statistical property of light whose photon number uncertainty is smaller than that in the coherent state of light. The photon-number state $|n\rangle$ and the squeezed state described below have such a property. They can be called non-classical photons, in which the photon number distribution takes a sub-Poisson distribution whose variance is smaller than that in a Poisson distribution [14].

The properties of the squeezed state, which are different from those of the coherent state, are described in the following. Coherent and hyper-coherent states of light correspond to classical sinusoidal waves whose fluctuation magnitudes have been limited by the quantum noise, i.e. spontaneous emission fluctuations or shot noise of the feedback loop. However, non-classical photons can be considered as exceeding this quantum noise limit in some meanings. Generally speaking, the product of the uncertainty of measurement of two Hermite-conjugate physical quantities, X_1 and X_2, can not exceed the limit imposed by the uncertainty principle, which has been described in Eq.(1-61). In the case of a coherent state, the minimum uncertainty condition given by Eq.(1-66) is realized, i.e. the equality of Eq.(1-61) holds and, furthermore, $\Delta X_1 = \Delta X_2$. However, the state in which $\Delta X_1 < \Delta X_2$ or $\Delta X_1 > \Delta X_2$ can be realized as long as this minimum uncertainty is maintained because the uncertainty principle is not violated. Realizing one of these situations is called squeezing. The state of the physical quantity obtained by this procedure is called a squeezed state. For example, if a variation in the optical phase is utilized for some optical systems, the uncertainty of the optical power does not have to be low. In this case, the condition $\Delta\phi < 1/\sqrt{2}$ and $\Delta n > 1/\sqrt{2}$ in Eq.(1-68) should be useful for this system. This condition corresponds to a squeezed state in which the phase fluctuation is low. If this phase-squeezing is realized for the frequency control in Chapter 5, a very low FM noise laser beyond the quantum noise limit can be developed. Furthermore, the measurement accuracy of the gravitational wave detector in 7.5(5) can be improved if a phase-squeezed light is employed. It is also possible to realize a photon number squeezed state, i.e. $\Delta\phi > 1/\sqrt{2}$ and $\Delta n < 1/\sqrt{2}$, by which the performances of a long-distance optical communication system making use of power modulation or photon counting could be improved. However, a careful treatment of squeezed light should be required since the uncertainty ΔX_2 of the counter-part physical quantity increases if that of the relevant physical quantity ΔX_1 is decreased. In this meaning, the characteristics of the squeezed state are different from those of coherent or hyper-coherent states.

Although the squeezed state can be realized in all physical systems as long as they are described by an harmonic oscillator model, recent research activities have been concentrated on light. To obtain a squeezed state, the values of ΔX_1 and ΔX_2

should be controlled by utilizing a pair of photons which are quantum mechanically correlated to each other. Thus, to obtain squeezed light, a light emission process has to be used, in which two mutually correlated photons are generated. In this meaning, it can be pointed out that the concept of squeezing is close to that of the photon pair in the phase conjugated wave generation described in 6.3.2. For example, a squeezed state can be realized if the Hamiltonian representing the photon generation process has the following form:

$$\mathcal{H} = \hbar\left(vaa + v^* a^+ a^+ \right). \tag{8-4}$$

Since most optical processes having this kind of Hamiltonian are nonlinear optical processes, a parametric amplification due to the second-order nonlinear susceptibility and four-wave mixing due to the third-order nonlinear susceptibility have been used. Prior to the generation and detection of squeezed light, several problems should be solved. The principal problems are: (1) Using an optical process which exhibits a large nonlinearity. (2) Additional optical processes such as absorption and spontaneous emission should be removed since squeezed light is easily destroyed by these processes. (3) Additional noises should be removed since a very low magnitude of fluctuations below the quantum noise is measured. It is effective to observe the beat signal between the relevant phase component of the measured light and the reference local light by the homodyne or heterodyne techniques in order to measure directly the fluctuations of the conjugate quantities ΔX_1 and ΔX_2.

Since the complicated experimental conditions presented above are required, it is not easy to obtain the squeezed light. However, after the successful experiments were reported in 1985 [15], several research groups succeeded in performing these experiments. The first experiments have employed a four-wave mixing process by using the Na atomic beam as a nonlinear medium, in which dye lasers were used as light sources [16]. The largest value of the noise suppression reported so far was realized by using a parametric process [16]. As shown in Fig. 8-6(a), the second harmonic wave from a Nd : YAG laser was used as a pumping wave (angular frequency; v_p) to obtain a parametric oscillation by a MgO : $LiNbO_3$ crystal in the cavity. By using this experimental setup, squeezed light of an angular frequency $v_p/2$ generated by the degenerated parametric oscillation was detected [15]. In order to reduce additional noises, several experimental techniques were employed, i.e. a parametric oscillator was operated below the threshold level, and the pumping wave was used as a local light in the balanced homodyne detection using an optical balanced detector. Figure 8-6(b) shows the obtained experimental result. It is found that the noise level was reduced to 50% of that of the quantum noise limit by adjusting the phase of the local light. Although a 50% squeezing has not yet been sufficient for the practical application systems and has not yet been superior to the applicability of hyper-coherent light to these systems, squeezing as large as 10 dB can be expected by using a higher quality nonlinear optical medium. The future problem remaining to be solved might be to find how to use this squeezed light in optical systems.

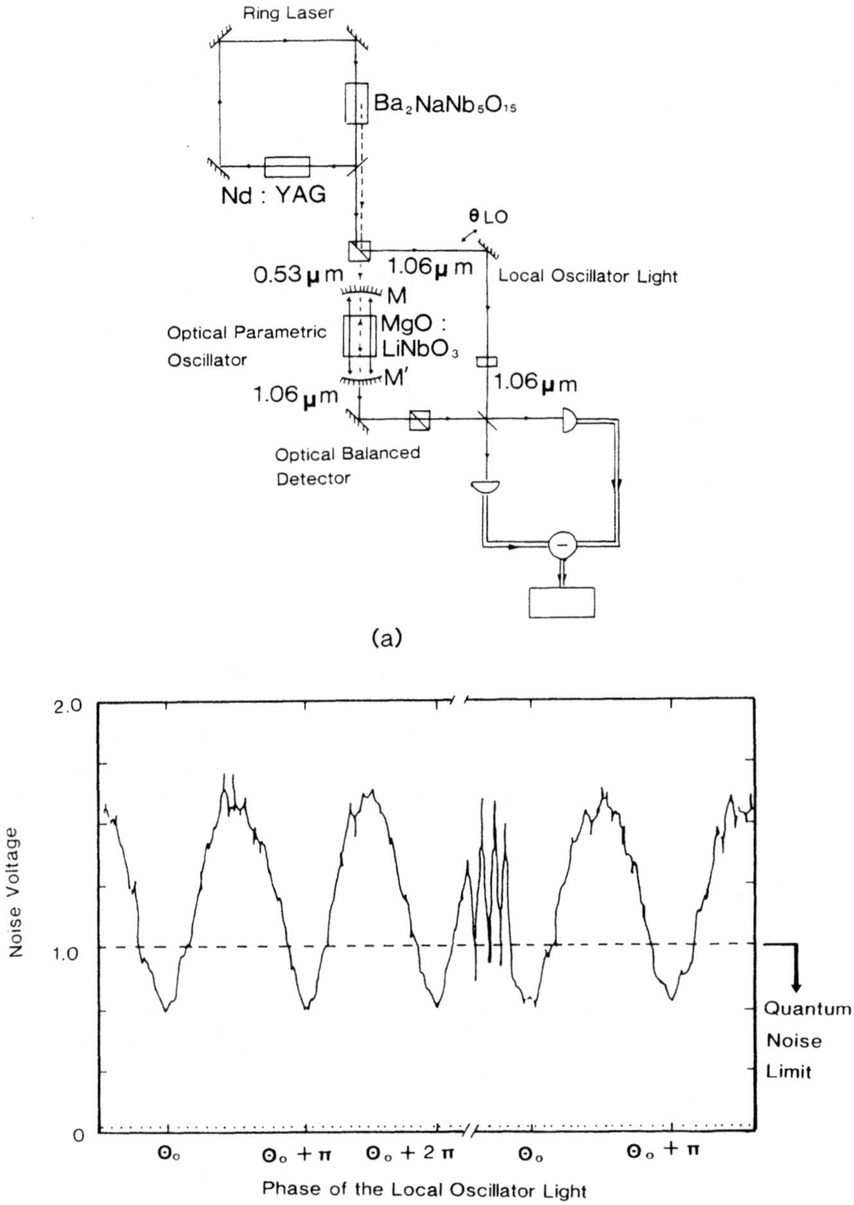

(a)

(b)

Fig. 8.6. Generation of a squeezed light by the optical parametric process (after Wu, Kimble, Hall, and Wu [16]). (a) Setup of a degenerated parametric amplifier and a balanced detector composed of a ring cavity YAG laser and a nonlinear crystal. (b) Measured noise magnitude. The broken line represents the magnitude of the quantum noise corresponding to the vacuum fluctuation.

8.3.2 Quantum nondemolition measurements

When a physical quantity is measured, the measurement process itself can perturb the condition of the physical quantity. For example, if the position of the free particle is measured with an error of Δq, this measurement induces a measurement error Δp in the momentum, which is given by Eq.(1-62) (i.e. $\Delta p \geq (\hbar/2\Delta q)$). This means that velocity fluctuations in the particle are induced by this measurement, which are as large as $\Delta v = \Delta p/M$, where M is the mass of the particle. As a result of these velocity fluctuations, the error of the position measured at time t after the initial measurement is $\Delta p \geq \sqrt{\hbar t / 2M}$. The magnitude of this additional error is sufficiently large not to be neglected, e.g. the measurement of the mirror position in the gravitational wave detection described in 7.5(5). To solve this problem, a new method has been proposed for gravitational wave detection [17]; it has been called quantum nondemolition measurement (QND measurement) [18]. As shown in Fig. 8-7, this is a method of measurement without inducing any change in the measured physical quantity. It can be easily found by the following discussion that this method is basically possible: In contrast to the above-mentioned measurement procedures, the measurement of momentum is discussed. If momentum is measured with an error of Δp, the minimum measurement error of the position Δq is given by Eq.(1-62). However, this error Δq does not affect the measurement error of the succeeding measurement of the momentum since no change in particle velocity has been induced by Δq. Thus, this procedure is the QND measurement for the momentum. The physical quantity for which the QND measurement is possible is called QND variable. The reason why the momentum p is the QND variable is that the Hamiltonian $\mathcal{H}$ ($\equiv p^2/2M$) of the free particle commutes with the momentum operator p. Since the Hamiltonian of the light field given in 1.2 has the same form

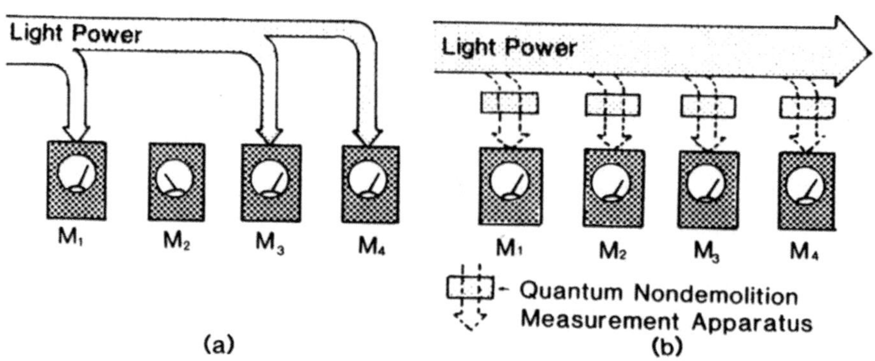

Fig. 8.7. Measurement of informations transmitted by light. (a) Common measurement. When the information is indicated by the meters ($M_1 \sim M_4$), a part of power must be dissipated for the measurements and lead the loss of the transmitted light power. (b) Quantum nondemolition measurement. Information can be measured without a power dissipation, therefore there is no power loss after the measurement.

as that of a harmonic oscillator, the QND variables are the photon number n. Furthermore, $a + a^+$ and $a - a^+$ can be the QND variables corresponding to the sine and cosine components of the classical electrical field amplitude, respectively.

Studies on QND measurements have been carried out in connection with those on the squeezed state since the generation of a quantum mechanical correlation by using nonlinear optical processes is commonly utilized for both types of research. Since research on the squeezed state was focused on generating a quantum mechanical correlation, low losses in nonlinear optical media have been allowable. As shown in Fig. 8-6, squeezed light has actually been observed. However, QND measurement is more difficult than the generation of a squeezed state because it is essential to avoid destroying the state of light by losses in nonlinear optical media. However, if QND measurements are realized, the measurement sensitivity of gravitational wave detection can be improved. As another example of applications, detection by tapping and demodulating the transmitted signal becomes possible without loosing its power in communication systems, and the transmission capacity can be dramatically increased.

8.4 Control and Manipulation of Atoms and Photons

Resonance frequencies in low pressure atomic vapor could be slightly shifted from an intrinsic Bohr frequency due to atomic collision, second-order Doppler effect, thermal motion, and so on. The preparation of new atomic states has been tried by controlling atomic motions to etiminate this shift in order to use atoms for an ultrahigh precision frequency standard, ultrahigh resolution laser spectroscopy and basic studies on quantum optics. These states can be obtained by the techniques of laser cooling and ion trap described in 7.5(3).

On the other hand, the techniques of controlling and manipulating photons could correspond to the cavity QED and photon localization which have been described in 5.4. and 6.3.3. As an application of cavity QED, the interaction between one atom and a high Q single mode cavity is described in the following [19, 20]. Such a situation implies that the vacuum has only one mode which interacts with the atom. In the case when a single mode light field of an angular frequency ν_q interacts with one atom, the Hamiltonian of this system can be expressed as

$$\mathcal{H} = \hbar\nu_q S^z + \hbar\nu_q a^+ a + \hbar g\left(S^+ a + S^- a^+\right), \qquad (8-5)$$

where S^z, S^+ and S^- are spin-operators for the atom, g is a coupling constant of the interaction between an atom and the vacuum field. If only the two cases: [(1) the atom is in the upper state $|a\rangle$ and no photon $|0\rangle$ or (2) the atom is in the lower state $|b\rangle$ and one photon $|1\rangle$] are considered, the state of this system can be expressed as

$$|\psi\rangle = c_a|a,0\rangle + c_b|b,1\rangle. \qquad (8-6)$$

By using the Hamiltonian of Eq.(8-5), the coefficients c_a and c_b of this expression can be given by

$$\left.\begin{aligned} c_a &= -i\sin(gt) \\ c_b &= \cos(gt) \end{aligned}\right\} \qquad (8-7)$$

The coupling coefficient g is expressed as [23]

$$g = -\frac{p}{\hbar}\sqrt{\frac{\hbar \nu_q}{\varepsilon_0 V}} \qquad (8-8)$$

where p is the transition dipole moment and V is the cavity volume. It is seen from Eq.(8-7) that the atom can move back and forth between the upper and lower states with an constant time period of $T(=2\pi/|g|)$ and consequently, the photon number also changes between 0 and 1 with this period. The discussion given above is similar to the previous formulation by Eqs.(2-18) ~ (2-29) of 2.2. by fixing $\omega = \nu$. However, in the previous case, the oscillation does not take place if $E_{x0} = 0$ because the Rabi angular frequency μ becomes 0. On the other hand, the g always takes non-zero value in the present system, i.e. an interaction between the atom and field can take place, which is called the Rabi oscillation induced by vacuum fluctuation. By utilizing this phenomenon, it can be expected that a very low threshold laser is possible. Although it is not straightforward to demonstrate this phenomenon experimentally in the optical frequency region, it has been already demonstrated in the microwave frequency region by using a Rydberg atom and superconductor cavity [21]. A small number of sodium (Na) and a high-Q Fabry-Perot interferometer have been utilized recently to demonstrate this phenomenon in the optical frequency region. As has been shown by Fig. 7-30, spectral splitting of the spontaneous emission has been observed, which was due to the Rabi oscillation in vacuum [20].

On the other hand, a method of realizing the correlation between the spontaneous emission fluctuations for the two lasing modes has been proposed [22]. It has not been believed that this kind of correlation would exist because the spontaneous emission can be interpreted as the stimulated emission triggered by the zero-point fluctuations of vacuum. However, if the upper levels of atom are consisted of more than two level, this correlation can be realized by realizing a quantum correlation between those upper levels by an external modulation. This correlation has been experimentally demonstrated for a gas laser [23] and a semiconductor laser [24]. Confusion between this correlation (being called as a correlated spontaneous emission) and a conventional phenomenon of active mode locking should be avoided because the characteristics of a part of correlated spontaneous emission is similar to those of the active mode locking. A method of accurate evaluation of the correlation has been proposed by accurately measuring the Allan variance of the short-term fluctuations of the phase difference between the two spontaneous emissions [25]. An application of this correlated spontaneous emission to the long base line interferometer for the gravitational wave detection has been also proposed [25].

Although theoretical studies of realizing a novel phenomenon and its application by controlling or manipulating the atoms and photons have been carried out for a long time, experimental studies have been drastically progressed only recently, which are attributed to the progresses in experimental techniques of laser control of chapter 6 and the techniques of material and device fabrications. On the other hand, it can be expected that the results of these studies can be applied back to the techniques of laser control, novel material and/or device fabrications. Studies of controls and manipulations of single atom or photon, and their interaction have been actively carried out recently. As the next step of the interaction between the single atom and single photon, a possibility of the photon-photon scattering of the intense laser radiation in vacuum has been predicted [26]. Furthermore, it has been proposed by this prediction that a second harmonic generation is also possible in a D.C. magnetized vacuum due to broken symmetry of the interaction [27]. These studies of quantum optics under an ultimate condition could establish a novel basis of physics, and open a possibility of progress in the coherent quantum optics and technology.

8.5 High Power Laser and Optical Energy Storage

High quality lightwave can be generated by a laser and its modulation, transmission, and detection can be also possible by using electro-optical materials, optical fibers, and photo-diodes, respectively. Optical communication and optical computing systems have been developed by using these techniques. However, efficient utilizations of optical energy have not yet been fully developed because the method of optical energy storage has not yet been found. This can be because a photon energy $\hbar\nu$ is low, and the spatial photon localization is not straightforward. An example of optical energy utilization can be laser fusion system. A progress has been made toward the break-even (a situation at which the output energy due to the laser fusion is equal to the input energy) by using high power lasers. If the break-even is achieved, further applications systems can be realized. For example, interplanetary rocket propulsion by laser fusion has been proposed [28].

If the optical energy from these high power lasers can be stored in an optical container and can be extracted from it, it can be expected that a wide variety of applications is possible, e.g., motive power, heating, dynamo, actuators [29], digital, logic devices, micro-robots, and micro-manipulators. Although the technique of the cavity QED of 8.5 can provide an example of optical energy storage, stored energy is still low. To realize practical optical power storage, a kind of power container such as a capacitor in the electric circuit, is required. A super-high finesse Fabry-Perot cavity corresponds to one of these containers, however, its storage time, i.e. a photon lifetime, is still shorter than several second. On the other hand, a low loss superconductor cavity has been already developed in the microwave frequency region [30]. Such a high quality cavity should be developed in the optical frequency region. To realize this kind of optical energy storage, basic technologies given by 8.4. and material and/or device fabrication techniques will play essential roles. If the light can be used not only for information transmission, but also for energy transmission, the era of coherent quantum optics and technology can really begin.

8.6 Summary

For the conventional design of high quality lasers, atomic motions have been controlled. Novel light sources can be expected in the future by controlling and manipulating the single atom and/or photon, and by utilizing the non-classical photons. Today looks like a transient period toward the new era of quantum optics. The scientists in this field are searching a direction toward this new era.

[References]

1) G. J. Dixon, C. E. Tanner, and C. E. Wieman, "432-nm source based on efficient second-harmonic generation of GaAlAs diode-laser radiation in a self-locking external resonant cavity", *Opt. Lett.*, Vol. 14 (1989) pp. 731–733.

2) T. Senoh, Y. Fujino, Y. Tanabe, M. Ohtsu, and K. Nakagawa, "Direct Modulation of Stable Blue Radiation from Frequency Doubled GaAlAs Laser Diode by Using EO Effect in $KNbO_3$ Nonlinear Crystal", *Proc. Conference on Lasers and Electro-Optics (CLEO'91)*, May 1991, Baltimore (Optical Soc. Am., Washington, D.C., 1991) CWA6.

3) W. J. Kozlovsky, C. D. Nabors, and R. L. Byer, "Efficient Second Harmonic Generation of a Diode-Laser-Pumped CW Nd : YAG Laser Using Monolithic MgO : $LiNbO_3$ External Resonant Cavities", *IEEE J. Quantum Electron.*, Vol. 24 (1988) pp. 913–919.

4) K. Kajikawa and H. Takezoe, "Nonlinear optics and material technology", *Optical and Electro-Optical Engineering Contact*, Vol. 27 (1989) pp. 405–415 (in Japanese).

5) D. Mehuys, M. Mittelstein, and A. Yariv, "(GaAl) As quantum well semiconductor lasers tunable over 105 nm with an external grating", *Proc. Conference on Lasers and Electro-Optics (CLEO'89)*, April 1989, Baltimore (Optical Soc. Am., Washington, D.C., 1989) FL4.

6) M. Ohtsu, "Progress toward Highly Coherent Semiconductor Lasers", *Proc. the 7th International Conference on Integrated Optics and Optical Fiber Communication (IOOC'89)*, July 1989, Kobe (Inst. Electron., Information, and Commun. Eng., Tokyo, 1989) 19A3-1.

7) H. Suzuki, I. Koshiishi, K. Nakagawa, and M. Ohtsu, "Frequency Control of Visible Semiconductor Laser", Extended Abstracts of the 50th Autumn Meeting of the Jpn. Soc. Appl. Phys., Sept. 1989, Fukuoka (Jpn. Soc. Appl. Phys., Tokyo, 1989) 29pZL7 (in Japanese).

8) C. R. Pollock, D. A. Jennings, F. R. Petersen, J. S. Wells, R. E. Drullinger, E. C. Beaty, and K. M. Evenson, "Direct frequency measurements of transitions at 520 THz (576 nm) in iodine and 260 THz (1.15 μm) in neon", *Opt. Lett.*, Vol. 8 (1983) pp. 133–135.

9) K. M. Evenson, J. S. Wells, F. R. Petersen, B. L. Danielson, and G. W. Day, "Accurate frequencies of molecular transitions used in laser stabilization; the 3.39 μm transition in CH_4 and the 9.33- and 10.18-μm transitions in CO_2", *Appl. Phys. Lett.*, Vol. 22 (1973) pp. 192–195.

10) Y. Enomoto and T. Murakami, "Optical Detector Using Oxide Superconductor", Extended Abstracts of the 36th Spring Meeting of the Jpn. Soc. Appl. Phys., Tokyo, April 1989 (Jpn. Soc. Appl. Phys., Tokyo, 1989) 2pZD4 (in Japanese).

11) M. Yamanishi, "Ultrafast Optical Processes through Virtual Charge Polarization in Quantum Well Structures", *Oyo Buturi*, Vol. 58 (1989) pp. 1696–1707 (in Japanese).

12) A. Yabe, Y. Taniguchi, H. Masuhara, and H. Matsuda, *Introduction to Organic Ultra-thin Films*, Baifukan Publishing, Tokyo, 1989 (in Japanese).

13) H. R. Telle, D. Meschede, and T. W. Hänsch, "Realization of a new concept for visible frequency division: phase locking of harmonic and sum frequencies", *Opt. Lett.*, Vol. 15 (1990) pp. 532–534.

14) T. Yajima, "Quantum effects of light", in *Quantum Electronics and New Technology*, ed. by Jpn. Soc. Phys., (Baifukan Publishing, Tokyo, 1987) (in Japanese).

15) R. E. Slusher, L. W. Hollberg, B. Yurke, J. C. Mertz, and J. F. Valley, "Observation of Squeezed States Generated by Four-Wave Mixing in an Optical Cavity", *Phys. Rev. Lett.*, Vol. 55 (1985) pp. 2409–2412.

16) L. A. Wu, H. J. Kimble, J. L. Hall, and H. Wu, "Generation of Squeezed States by Parametric Down Conversion", *Phys. Rev. Lett.*, Vol. 57 (1986) pp. 2520–2523.

17) V. B. Braginsky, Y. I. Vorontsov, and K. S. Thorne, "Quantum Nondemolition Measurements", *Science*, Vol. 209 (1980) pp. 547–557.

18) N. Imoto and S. Saito, "Quantum Nondemolition Measurement of Light", *Oyo Buturi*, Vol. 57 (1988) pp. 1535–1540 (in Japanese).

19) G. S. Agarwal, "Vacuum-Field Rabi Splittings in Microwave Absorption by Rydberg Atoms in a Cavity", *Phys. Rev. Lett.*, Vol. 53 (1984) pp. 1732–1734.

20) M. G. Raisen, R. J. Thompson, R. J. Brecha, H. J. Kimble, and H. J. Carmichael, "Normal-mode splitting and linewidth averaging for two-state atoms in an optical cavity", *Phys. Rev. Lett.*, Vol. 63 (1989) pp. 240–243.

21) D. Meschede, H. Walther, and G. Muller, "One-atom maser", *Phys. Rev. Lett.*, Vol. 54 (1985) pp. 551–554.

22) M. O. Scully, "Correlated Spontaneous-Emission Lasers: Quenching of Quantum Fluctuations in the Relative Phase Angle", *Phys. Rev. Lett.*, Vol. 55 (1985) pp. 2802–2805.

23) P. E. Toschek and J. L. Hall, "Noise reduction in the relative phase of two lasers", Tech. Digest of XV International Quantum Electronics Conference (IQEC'87), Baltimore, April-May 1987 (Opt. Soc. Am., Washington, D.C, 1987) WDD2.

24) M. Ohtsu and K. -Y. Liou, "Correlated spontaneous emission between two longitudinal modes in an extended-cavity semiconductor laser", *Appl. Phys. Lett.*, Vol. 52 (1988) pp. 10–12.

25) M. Winters, Joint Institute for Laboatory Astrophysics, Unov. Colorado, private communication, 1989.

26) W. Heisenberg and H. Euler, "Fogerungen aus der Diracschen Theorie des Positions", *Z. Phys.*, Vol. 98 (1936) pp. 714–732.

27) Y. J. Ding and A. E. Kaplan, "Nonlinear Magneto-Optics of Vacuum: Second-Harmonics Generation", *Phys. Rev. Lett.*, Vol. 63 (1989) pp. 2725–2728.

28) H. Nakajima, "Is a Laser Fusion Rocket Possible?", *Parity*, Vol. 4 (1989) pp. 69–71 (in Japanese), and references therein, e.g. R. Hyde, L. Wood, J. Nuckolls, "Prospects for Rocket Propulsion with Laser-Induced Fusion Microexplosions", AIAA Paper No. 72-1063 (1972).

29) T. Morikawa, "Expectation for Optical Actuators", *Proc. 1st Meeting on Lightwave Sensing Technol., Tokyo, June 1988* (Lightwave Sensing Technol. Res. Group, Jpn. Soc. Appl. Phys., Tokyo, 1988) LST1-5 (in Japanese).

30) S. R. Stein and J. P. Tuneaure, "Superconducting-Cavity Stabilized Oscillators with Improved Frequency Stability", *Proc. IEEE*, Vol. 63 (1975) pp. 1249–1250.

Appendix: Rate Equations and Relaxation Oscillation

In some cases, the characteristics of the temporal variation of the laser power, i.e., the photon numbers, are discussed, in which the phase of light field could be neglected. This appendix is devoted to derive the equations to describe the temporal variations of photons [1].

Since the phase of the electric field of light is neglected, one can assume that $\nu_{ql} = \nu_q = \omega$ in Eq.(2-75) to get

$$\frac{\mathrm{d}E}{\mathrm{d}t} = -\frac{\omega}{2Q}E + \frac{i\omega}{2\varepsilon_0}P \qquad (A-1)$$

where the subscript q was removed in order to discuss the case of a single-longitudinal mode operation for simplicity. The photon number n can be derived from the optical energy $(\varepsilon_0/2)|E|^2$ per unit volume, which is expressed as

$$n = \frac{\varepsilon_0}{2\hbar\omega}|E|^2 \qquad (A-2)$$

From Eq.(A-1), one derives

$$\frac{\mathrm{d}n}{\mathrm{d}t} = -\frac{\omega}{Q}n + \frac{i}{4\hbar}\left(PE^* - P^*E\right) \qquad (A-3)$$

Since the optical phase is neglected, one can fix the value of U_q to unity in Eqs.(2-56) and (2-58). Then, one obtains

$$P = -\frac{i|p|^2}{\hbar\gamma}E\Delta\rho \qquad (A-4)$$

where

$$\Delta\rho \equiv \rho_{aa} - \rho_{bb} \qquad (A-5)$$

is the population inversion. Substituting Eq.(A-4) into Eq.(A-3), one gets

$$\frac{\mathrm{d}n}{\mathrm{d}t} = -\frac{1}{\tau_p}n + \frac{\omega|p|^2}{\hbar\varepsilon_0\gamma}n \cdot \Delta\rho \qquad (A-6)$$

where Eqs.(2-16) and (2-17) were used. On the other hand, from Eqs.(2-46) and (2-47), the following expression can be obtained.

$$\frac{\mathrm{d}\Delta\rho}{\mathrm{d}t} = -\frac{1}{\tau_s}\left(\Delta\rho - \Delta\rho^0\right) - \frac{2\omega|p|^2}{\hbar\varepsilon_0\gamma}n \cdot \Delta\rho \qquad (A-7)$$

where it was assumed that $\gamma_a = \gamma_b$ $(= 1/\tau_s)$. τ_s is the spontaneous emission lifetime, i.e. time until the atom staying in the upper-energy level radiates spontaneous emission. $\Delta\rho^0$ is defined by

$$\Delta\rho^0 \equiv \rho_{aa}^0 - \rho_{bb}^0 \qquad (A-8)$$

The quantity

$$B_0 \equiv \frac{\omega|p|^2}{\hbar\varepsilon_0\gamma} \qquad (A-9)$$

is proportional to Einstein's B coefficient in Eq.(2-30), which represents the gain via stimulated emission. Then, Eqs. (A-6) and (A-7) are reduced to

$$\frac{\mathrm{d}n}{\mathrm{d}t} = -\frac{1}{\tau_p}n + B_0 \cdot n \cdot \Delta\rho \qquad (A-10)$$

and

$$\frac{\mathrm{d}\Delta\rho}{\mathrm{d}t} = -\frac{1}{\tau_s}\left(\Delta\rho - \Delta\rho^0\right) - 2B_0 \cdot n \cdot \Delta\rho \qquad (A-11)$$

These are the coupled nonlinear equations used to describe the relationship between photon numbers and population inversion and they are called rate equations. The second term in Eq.(A-11) is two times the second term in Eq.(A-10), which means that one photon increase due to stimulated emission will lead to two atoms decreases in population inversion.

In the derivation of these rate equations, the temporal variations of E, ρ_{ab} and $\rho_{aa} - \rho_{bb}$ were assumed to be much smaller than the constant γ (, i.e., $d/dt << \gamma$). According to these assumptions, it is confirmed that these rate equations are valid only when the following inequalities are satisfied:

$$\left| \frac{1}{n}\frac{dn}{dt} + \frac{2}{\Delta\rho}\frac{d\Delta\rho}{dt} \right| << 2\gamma$$

$$\left| \frac{1}{n}\frac{dn}{dt} \right| << 2\gamma \qquad (A-12)$$

$$\left| \frac{1}{\Delta\rho}\frac{d\Delta\rho}{dt} \right| << \gamma$$

Then, one can conclude that

(1) Rate equations are valid for a steady-state in which $d/dt = 0$.

(2) Rate equations are valid for the relaxation oscillation because the temporal variations of n and $\Delta\rho$ compensate with each other.

(3) Rate equations will become invalid possibly in the process of a giant pulse generation via a Q-switch because n and $\Delta\rho$ may vary simultaneously.

(4) Rate equations are invalid for a gas laser even in its steady-state operation because the perturbation exerted on each atom is different between each other due to the Doppler effect caused by the thermal motion of atoms.

As a first step, the characteristics of a steady-state are discussed by using Eqs.(A-10) and (A-11). By fixing $d/dt = 0$, one gets

$$\Delta\rho^s = B_0 / \tau_p (\equiv \Delta\rho_{\text{th}}) \qquad (A-13)$$

and

$$n^s = \frac{1}{2B_0\tau_s}\left(\frac{\Delta\rho^0}{\Delta\rho_{\text{th}}} - 1\right) \qquad (A-14)$$

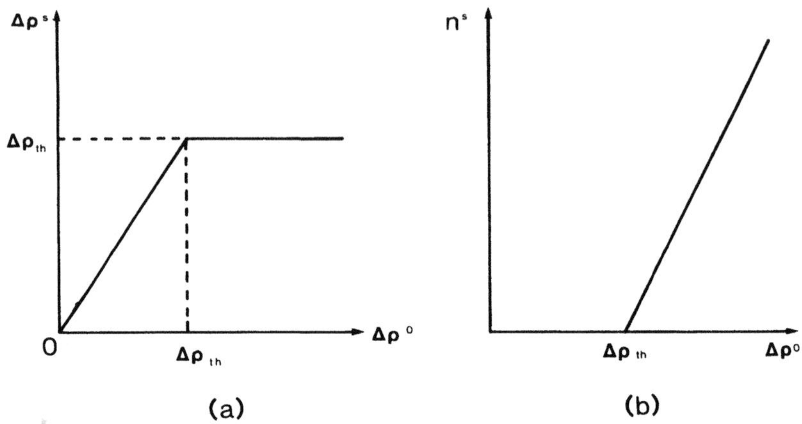

Fig. A.1. Variations of the steady state values of the population inversion $\Delta\rho^s$(a) and the photon number n^s(b) with the increase of the energy supplied from the power source (, which is proportional to $\Delta\rho^0$). $\Delta\rho_{th}$ represents the threshold value of the population inversion required for laser oscillation.

As shown in Fig. A-1, these expressions mean that the population inversion $\Delta\rho$ is increased up to a threshold value $\Delta\rho_{th}$ by increasing $\Delta\rho^0$, i.e. by increasing the energy supplied from the power source.

For $\Delta\rho^0 > \Delta\rho_{th}$, the photon number can increase linearly with an increase in $\Delta\rho^0$ as is given by Eq.(A-14). These characteristics correspond to those given in Fig. 2.7. In this case, the population inversion is clipped to its threshold value, as is shown in Eq.(A-13).

As a second step, the discussion is focused on the case when population inversion shows a slight temporal variation induced by a variation in supplied energy $\Delta\rho^0 + \delta\Delta\rho^0(t)$. Equations (A-10) and (A-11) are linearized by adding the small fractions $\delta\Delta\rho(t)$ and $\delta n(t)$ to the steady-state values of Eqs.(A-13) and (A-14). They are expressed as

$$\frac{\mathrm{d}(\delta n)}{\mathrm{d}t} = -\frac{1}{\tau_p}\cdot\delta n + B_0\left(n^s\delta\Delta\rho + \Delta\rho^s\delta n\right) \qquad (A-15)$$

and

$$\frac{\mathrm{d}(\delta\rho)}{\mathrm{d}t} = -\frac{1}{\tau_s}\left(\delta\Delta\rho - \Delta\rho^0\right) - 2B_0\left(n^s\delta\Delta\rho + \Delta\rho^s\delta n\right) \qquad (A-16)$$

By combining these equations, one obtains a linear differential equation for $\delta n(t)$, which is given by

$$\frac{\mathrm{d}^2(\delta\rho)}{\mathrm{d}t^2}+\frac{1}{T}\frac{\mathrm{d}(\delta n)}{\mathrm{d}t}+\Omega^2\delta n=\frac{B_0 n^s}{\tau_s}\delta\Delta\rho^s \qquad (A-17)$$

where

$$\frac{1}{T}\equiv\frac{1}{\tau_s}\frac{\Delta\rho^s}{\Delta\rho_{\mathrm{th}}} \qquad (A-18)$$

and

$$\Omega^2\equiv\frac{1}{\tau_s\tau_p}\left(2\frac{\Delta\rho^s}{\Delta\rho_{\mathrm{th}}}-1\right) \qquad (A-19)$$

It is easily found that the solution of this differential equation can be expressed as

$$\delta n\propto n^s\left[1-\exp\left\{\left(-\frac{1}{T}+i\Omega\right)t\right\}\right] \qquad (A-20)$$

which shows a form of relaxation oscillation with a frequency

$$f_r\equiv\frac{\Omega}{2\pi}=\frac{1}{2\pi\sqrt{\tau_s\tau_p}}\sqrt{2\frac{\Delta\rho^s}{\Delta\rho_{\mathrm{th}}}-1} \qquad (A-21)$$

This characteristic frequency is called the relaxation oscillation frequency.

[Reference]

1) K. Shimoda, T. Yajima, Y. Ueda, T. Shimizu, and T. Kasuya, *Quantum Electronics*, Shokabo Publishing, Tokyo, 1972 (in Japanese).

INDEX

GPSR Compliance

The European Union's (EU) General Product Safety Regulation (GPSR) is a set of rules that requires consumer products to be safe and our obligations to ensure this.

If you have any concerns about our products, you can contact us on ProductSafety@springernature.com

In case Publisher is established outside the EU, the EU authorized representative is:

Springer Nature Customer Service Center GmbH
Europaplatz 3
69115 Heidelberg, Germany

Batch number: 10607320

Printed by Printforce, the Netherlands